W9-BUL-570

HONDURAS
GUATEMALA
EL SALVADOR
NICARAGUA
COSTA RICA
COLOMBIA
ECUADOR
PERU
BOLIVIA
CHILE

NON-GOVERNMENTAL ORGANIZATIONS AND THE STATE IN LATIN AMERICA

This book is in the Non-Governmental Organizations Series, which comprises three regional studies and a synthetic overview volume. Each of the titles presents detailed empirical insights into the work of Non-Governmental Organizations in agriculture. The case material is set within the context of NGOs' relations with the state and their contribution to democratization and the consolidation of rural civil society.

The books are written at a time of massively increased funding for NGOs, of increasing fiscal stringency in the public sector attributable to structural adjustment, and of growing pressures on NGOs from governments and donors to rethink their roles.

Against the background of a broad review of institutional activity at the grassroots, each book explores specific questions concerning the work of NGOs in agricultural development:

- How good/bad are NGOs at promoting technological innovation and addressing constraints to change in peasant culture?
- How effective are NGOs at strengthening grassroots/local organizations?
- How do/will donor pressures influence NGOs and their links to the state?

Eschewing populist acclaim for NGOs, these books draw on a large volume of new empirical material to provide a comprehensive review and critique of the performance and capabilities of these organizations.

These books will find a wide readership among students, academics and practitioners in the field.

Non-Governmental Organizations and the State in Latin America is by **Anthony Bebbington**, Overseas Development Institute, **Graham Thiele**, British Tropical Agriculture Mission and **Penelope Davies**, British Tropical Agriculture Mission with Martin Prager, CELATER and Hernando Riveros, CELATER.

NON-GOVERNMENTAL ORGANIZATIONS SERIES

Co-ordinated by the **Overseas Development Institute**

The titles in the series are:

RELUCTANT PARTNERS?
Non-Governmental Organizations, the State and Sustainable Agricultural Development
John Farrington and Anthony Bebbington with Kate Wellard and David J. Lewis

NON-GOVERNMENTAL ORGANIZATIONS AND THE STATE IN LATIN AMERICA
Rethinking Roles in Sustainable Agricultural Development
Anthony Bebbington and Graham Thiele with Penelope Davies, Martin Prager and Hernando Riveros

NON-GOVERNMENTAL ORGANIZATIONS AND THE STATE IN AFRICA
Rethinking Roles in Sustainable Agricultural Development
Edited by Kate Wellard and James G. Copestake

NON-GOVERNMENTAL ORGANIZATIONS AND THE STATE IN ASIA
Rethinking Roles in Sustainable Agricultural Development
Edited by John Farrington and David J. Lewis with S. Satish and Aurea Miclat-Teves

NON-GOVERNMENTAL ORGANIZATIONS AND THE STATE IN LATIN AMERICA

Rethinking roles in sustainable agricultural development

Anthony Bebbington
and Graham Thiele
with
Penelope Davies, Martin Prager
and Hernando Riveros

London and New York

First published 1993
by Routledge
11 New Fetter Lane, London EC4P 4EE

Simultaneously published in the USA and Canada
by Routledge
29 West 35th Street, New York, NY 10001

Typeset in Garamond by
Witwell Limited, Southport
Printed and bound in Great Britain by
Biddles Ltd, Guildford and King's Lynn

British Library Catalolguing in Publication Data
A catalogue record for this book is available from the British Library

Library of Congress Cataloging in Publication Data
has been applied for

ISBN 0-415-08845-3
ISBN 0-415-08846-1 (pbk)

CONTENTS

FIGURES

TABLES

BOXES

FOREWORD

As Latin America finally emerges from the depth of the debt crisis, and as many countries again achieve positive per capita growth rates, the unresolved issue of equitable growth is coming back with a vengeance. During the 1960s and 1970s, recriminations concerning the unequalizing effects of rapid growth had been muffled by authoritarian regimes. During the 1980s, attention was diverted to the management of stabilization and adjustment programmes to reactivate the economies, accompanied by social compensation programmes to manage the political feasibility of reforms which had the potential of creating long-term social gains but had definite short-term costs. In the 1990s, as growth is resuming, the same structural features which had created inequality in the 1960s and 1970s are still in place, often futher reinforced by asset concentration as a result of extensive privatization programmes and the weakening of the middle class. There are, however, two significant differences with the past that now makes this growth style difficult to sustain politically. One is the most countries have shifted to democratic regimes as the military were delegitimized by the debt crisis when they could no longer compensate with growth the social costs of inequality. The other is that the grassroots have been massively organized as populations sought, through decentralized organizations, substitutes for what neither the market nor the state could offer them: well-being and democratic rights. With democratic regimes and a strongly organized civil society, unequal growth is no longer as politically easy to impose. Sustainability of the current economic recoveries very much depends on the ability of governments to engineer their long-term political feasibility.

In the quest for a politically feasible growth path, two approaches are being confronted. One is to seek rapid growth in the existing structure and to provide export compensations to ensure political stability. This approach was effective to manage recovery in the 1980s, as exemplified by Mexico's mix of trade liberalization (NAFTA) and solidarity (PRONASOL), but it is difficult to sustain politically over time and is highly costly. The other is to recognize that inequality is, in the long run, inefficient and that it is better to seek an integrated approach able to deliver simultaneously growth, welfare and political sustainability. This approach requires increasing the productive participation of the middle and poor sectors, and relying on an active state acting as a guardian of competitiveness and equity, and watching over market failures and strategic opportunities. The single most important lesson provided to Latin America by the Asian NICs is the demonstration that equity reduces resource wastage by mobilizing the productive potential of the mass of the population, motivates work and creativity, and reduces public expenditures on security and welfare. Mobilization of the productive potential of the poor in turn requires assisting their organizations help increase productivity, reduce transactions costs in accessing markets and act as lobbies to influence policy definition and gain greater access to public goods.

While the crisis displaced poverty to the cities, the recovery of growth is benefiting differentially the urban sector as it did in the 1960s and 1970s, reclocating extreme poverty in the rural areas. In addition, much urban poverty in marginal settlements is only the symptom of failed rural development inducing excessively rapid migration. As a consequence, the agenda for equitable growth very much requires delivering jobs and incomes in the rural areas.

With the return to democracy, the proliferation of popular movements, a scaled-down state no longer able to be the main engine of investment and welfare and thus seeking new partnerships, and the culmination of many blatant distortions in price policy against agriculture in particular, an opportunity exists to address anew the question of equitable growth by redefining the relationship between state, market and civil society. In particular, the possibility exists for a new partnership between grassroots organizations (GRO) and government. In assisting GROs, a vast array of NGOs, which also acted as substitutes for the state and/or the market from the 1960s throught the crisis, has emerged as a powerful instrument to help GROs develop strategies complementary to the state and the market. These grassroot support organizations offer a new instrumentality for a more competitive, sustainable, participatory and equitable development. This book addresses this challenge, placing NGOs in their broad economic, political and instrumental context, and identifies both the potential that collaboration between the state and NGO offers and the difficulties which it presents. More specifically, it addresses the potential for cooperation between NGOs and national-agricultural research and extension services (NARs) in delivering new sustainable technologies for smallholders. It does this by developing a number of case studies of NARs, NGOs, and NGO coordinating bodies, providing fascinating information that has to this date been highly dispersed, and extracting from it patterns that show both opportunities and weaknesses in state–NGO collaboration.

In addressing this important topic, Anthony Bebbington, Graham Thiele and their associates have placed themselves squarely at the center of the debate on how to define a new post-debt strategy for Latin America that integrates growth, equity, sustainability and democracy. They offer both a first class analysis, with a pragmatic balance in the identification of opportunities and difficulties, and a set of highly concrete recommendations on how to implement this strategy that will be invaluable to teachers, practitioners and donors in the fields of rural development and agricultural technology.

Alain de Janvry
University of California at Berkeley
February 1993

PREFACE

John Farrington

Interest in participation by the rural poor in the design and implementation of changes affecting their livelihoods, until recently a concern only of lobbyists on the fringe, is now moving to occupy centre stage in development debates. But practical experience with participatory approaches has so far been small-scale and localized. What organizational and institutional conditions need to be met if participatory approaches are to be implemented on a large scale? This book is one of a set of four that address this question.[1]

The book's focus is on agricultural change, particularly on how the types of technology and management practice necessary for sustainable improvement in agricultural productivity among small-scale, low-income farmers might best be developed. Institutionally, its focus is primarily on the work of non-governmental organizations (NGOs) in this sphere, but also on that of government research and extension services and, importantly, on the scope for closer interaction between the two.

These twin foci – on agriculture and on NGOs – are central to a number of policy concerns: the rural poor in many countries continue to rely for much of their livelihood on agriculture, yet renewable natural resources are coming under increasing population pressure. Government research institutes have found agricultural technologies harder to design for difficult rather than for well-endowed areas, and not all of the changes that have been adopted have proven technically or institutionally sustainable. At the same time, assistance to developing countries provided by and through NGOs is increasing rapidly and now amounts to almost one-fifth of net bilateral flows. Growing interest in NGOs is driven by perceptions of their role in democratic pluralism, by the hope that they might share some of the costs of providing development services and, significantly in the present context, by their perceived ability to reach the rural poor. This book is, above all, an empirical study: by focusing on their work in agriculture it attempts to move the debate about NGOs' potentials and limitations from the rhetorical to the concrete.

The book and its companion volumes bring together the findings of a research study initiated by the Agricultural Research and Extension Network of the Overseas Development Institute (ODI) in 1989. The study grew out of a background paper[2] for a workshop on farmer participation in agricultural research at the Institute of Development Studies, the University of Sussex.[3] The paper reviewed around 100 experiences of working with farmers in research and extension, many of them received from Network members.

During 1988 and 1989, we discussed our interest in the institutional prerequisites for large-scale implementation of these methods with practitioners in the field, many of whom expressed willingness to write up their experience in what has hitherto been a sparsely-documented field.

By 1991, documentation was in progress in eighteen countries covering Africa, Asia and South America, and this book is one of three continent-wide compilations of experiences. In addition to these books, efforts to make available the results of the Africa component of the study to policy-makers at national and international levels have been made through:

- the publication of case studies in the ODI Network;
- thematically-focused papers presented at international seminars (e.g. Farrington and Bebbington 1991; Bebbington and Farrington 1992);
- sponsorship of national workshops in Kenya and Zimbabwe;
- a regional workshop for West Africa, sponsored jointly with the International Institute for Environment and Development (IIED 1992).

The workshops in particular highlighted the extent to which entrenched attitudes, often perpetuated by lack of information, continue to act as a barrier to effective linkages between NGOs and the public sector; barriers that we hope this book and the broader story of which it is part have begun to break down.

NOTES

1 The other three are: *Reluctant Partners?: Non-Governmental Organizations, the State and Sustainable Agricultural Development* by John Farrington and Anthony Bebbington with Kate Wellard and David J. Lewis; *Non-Governmental Organizations and the State in Africa: Rethinking Roles in Sustainable Agricultural Development* edited by Kate Wellard and James G. Copestake; *Non-Governmental Organizations and the State in Asia: Rethinking Roles in Sustainable Agricultural Development* edited by John Farrington and David J. Lewis with S. Satish and Aurea Miclat-Teves.

2 Farrington and Martin 1987.

3 'Farmers and agricultural research: complementary methods' co-ordinated by Robert Chambers and held at the Institute of Development Studies, University of Sussex, in July 1987. For the edited proceedings, see Chambers *et al.* (eds) 1989.

ACKNOWLEDGEMENTS

Stephen Biggs has argued that nobody can claim sole responsibility for a new agricultural technology: there are multiple sources of agricultural innovation. Likewise there are multiple sources of intellectual production. Formal authorship is therefore less a statement about who contributed to a final written product than one about responsibility for the exact form and content taken by that final product. In the case of this book it will quickly become apparent to the reader that very many people have contributed to it. While it is we who are responsible for the tenor of the argument in the following pages, we gladly acknowledge the significant contributions that colleagues, friends and professional acquaintances have made to this study.

To introduce our most important acknowledgements, a word must be said about the genesis of this book. The study grew out of an experience in lowland Bolivia in which a public-sector research organization, the Centro de Investigación Agricola Tropical (CIAT), had begun to work with non-governmental organizations in agricultural development work with resource-poor farmers. This experience gave rise to a series of questions: what general lessons could be gained for agricultural development programmes elsewhere, how replicable was the model, and indeed was it desirable to replicate it at all? Out of these questions come this study, and similar ones in Asia and Africa. Along with the authors, John Farrington, who co-ordinated the whole research project from the Overseas Development Instititute, did initial spadework in seeing the project off the ground, and in developing the thinking behind it. As he was heavily involved in the study in Asia, he was unable to become involved in depth in the Latin American work. None the less, his intellectual, professional and personal contributions to the product you have before you have been very important.

As the research was initiated, the Columbian NGO, CELATER (the Latin American Centre for Rural Technology and Education) joined the team. At CELATER, Martin Prager took responsibility for the Columbian and Peruvian case material. The staff at CELATER, and particularly its Director, Enrique Castellanos, provided intellectual and logistical support throughout the research programme.

The research results were initially presented and discussed at a workshop held in Santa Cruz, Bolivia, in December 1991, hosted by CIAT and the NGO network in Santa Cruz; UNICRUZ. The support of CIAT, the British Tropical Agricultural Mission and UNICRUZ was total – they made the workshop happen. We would like to thank them all, and in particular Carlos Roca, José Quiroga, Roy Velez, Jonathon Wadsworth and John Wilkins.

Along the way conversations with many people have influenced this research. As we recognize them, we will inevitably omit more than we include, but special mention must be made of all the authors listed in Annex 1 who prepared the case studies on which the research is based. They gave their time not only to the documents they prepared, but also to interviews and conversations with the study co-ordinators who discussed with them many aspects of the work of non-governmental and governmental organizations, at times literally into the early hours of the morning. They also gave the co-ordinators friendship and logistical support as we worked alongside them in the preparation and revision of the manuscripts and in the collection of other information in their countries. They have been excellent colleagues and friends.

As the project, and then the manuscript, moved on we discussed its ideas with many people, and in particular we would like to recognize the comments, criticisms and support of the following: Stephen Biggs, Jutta Blauert, Alan Bojanic, Diego Bonifaz, Francisco Carrión, Tom Carroll, Enrique Castellanos, John Clark, Elana Dallas, John Farrington, Peter Gubbels, David Kaimowitz, Ben Kohl, David Lehmann, Shannon Mallory, Carlos Moreno, Harry Potter, Galo Ramón, Charlie Reilly, Carlos Roca, Octavio Sotomayor, Gillian Tett, Jorge Uquillas, Joachim Voss, Jonathon Wadsworth, Kate Wellard and Simon Zadek. At ODI Peter Ferguson's knowledge of the resources in the library was invaluable, and his assistance is warmly appreciated.

The research and the workshop were supported by the Overseas Development Administration (UK), the International Development Research Centre (Canada) and the Inter-American Foundation (USA), a truly international effort. At each of these institutions the people who supported us financially did so morally and intellectually as well: our thanks to all of them.

We would also like to express our gratitude to our respective employers, who supported us as we travelled, researched and wrote. The opinions expressed in the book, however, are entirely our own and do not reflect official views of any of the organizations for whom we work. Anthony Bebbington also wishes to recognize the support of the Centre of Latin American Studies at Cambridge University with whom he worked during the major part of this research.

The preparation, typing and formating of the manuscript was done at ODI under considerable pressure, and we are extremely grateful to Alison Saxby for her very skilful and patient contribution to the book. Elana Dallas's editing hand helped turn some tortuous prose into a far more accessible text. Carol

Whitwill, Kate Cumberland and Geraldine Healey also helped prepare the manuscript. Rachel Sieder helped with translation, and Richard Ketley and John Nelson with statistical information. The support of Tristan Palmer at Routledge was a source of reassurance as we worked under this pressure.

It may well be that many of these people who have helped see this book through to completion will share neither its conclusions nor its interpretations. Nevertheless, without them we would never have been able to present our argument in print in order to give them something with which to disagree. Our thanks.

ABBREVIATIONS

ACPH	Popular Cultural Action of Honduras (Honduras)
ADRO	Western Regional Rural Development Association (Honduras)
AGRARIA	AGRARIA: Food and Campesino Development (Chile)
AIPE	Association of Institutions for Popular Education (Bolivia)
ATD	Agricultural technology development
ATS	Agricultural technology system
BNF	National Agrarian Bank (Ecuador)
BTAM	British Tropical Agricultural Mission (Bolivia)
CAAP	Andean Centre for Popular Action (Ecuador)
CARE	American Cooperative for Remittances to the Exterior
CAS	Agricultural Union Centre (Bolivia)
CATIE	Research and Training Centre for Tropical Agronomy (Costa Rica)
CATT	Centre for the Adjustment and Transfer of Technology (INIA-Chile)
CCTA	Coordinating Commission for Andean Technology (Peru)
CDT	Working community (CIPCA-Bolivia)
CEDECO	Educational Corporation for Development (Costa Rica)
CEDRO	Rural Development Centre (El Salvador)
CEDLA	Centre for Labour and Agrarian Studies (Bolivia)
CEMAT	Mesoamerican Center for the Study of Appropriate Technology (Guatemala)
CEPA	Centre for Education and Agricultural Promotion (Nicaragua)
CEPAL	Economic Commission for Latin America (UN)
CEPLAES	Centre for Planning and Social Studies (Ecuador)
CESA-Bol	Centre for Agricultural Services (Bolivia)
CESA-Ec	Ecuadorian Centre for Agricultural Services (Ecuador)
CET	Centre for Education and Technology (Chile)

CGIAR	Consultative Group for International Agricultural Research
CIAT-Bol	Centre for Tropical Agricultural Research (Bolivia)
CIAT-Col	International Centre for Tropical Agriculture (Colombia)
CIDA	Interamerican Committee for Agricultural Development
CIDICCO	International Centre for Cover Crop Information and Documentation (Honduras)
CIED	Centre for Research, Education and Development (Peru)
CIMMYT	International Centre for Maize and Wheat Improvement (Mexico)
CIP	International Potato Centre (Peru)
CIPCA	Centre for Campesino Research and Development (Bolivia)
CIPRES	Centre for Research, Promotion and Rural and Social Development (Nicaragua)
CLADES	Latin American Consortium for Agroecology and Development (Latin America-Chile)
CNIGB	National Centre for Basic Grains Research (Nicaragua)
CONAF	National Forestry Corporation (Chile)
COOPEAGRO	Department for Agricultural Education Cooperation of El Ceibo (Bolivia)
CONAIE	Confederation of Indigenous Nationalities of Ecuador (Ecuador)
COPROALDE	Coordinator of NGOs with Alternative Development Projects (Costa Rica)
CORDECRUZ	Departmental Development Corporation of Santa Cruz (Bolivia)
COTESU	Swiss Technical Cooperation (Switzerland)
COTRISA	Wheat Marketing Corporation Ltd (Chile)
CUSO	Canadian University Services Overseas (Canada)
COSUDE	Swiss Development Cooperation (Switzerland)
CRI	Regional research centre
CRS	Catholic Relief Services (USA)
CVC	Autonomous Regional Corporation of the Cauca River Valley (Colombia)
DRI	Integrated Rural Development Project (Ecuador)
DTT	Department of Technology Transfer (CIAT-Bolivia)
EEC	European Economic Community
El CEIBO	The El Ceibo Regional Agricultural and Agro-industrial Cooperative Coordinating Committee (Bolivia)
FAO	United Nations Food and Agriculture Organization (UN)
FBU	United Brethren Foundation (Ecuador)

FEPP	Ecuadorian Fund for the Progress of the People (Ecuador)
FMLN	Farabundo Marti National Liberation Front (El Salvador)
FOSIS	Social Solidarity and Investment Fund (Chile)
FUNDAEC	Foundation for the Application and Teaching of Science (Colombia)
FUNDAGRO	The Foundation for Agricultural Development (Ecuador)
FUNDEAGRO	The Foundation for Agricultural Development (Peru)
GIA	Group for Agrarian Research (Chile)
GO	Government organization
GSO	Grassroots support organization
ICTA	Agricultural Science and Technology Institute (Guatemala)
IBTA	Bolivian Institute for Agricultural Technology (Bolivia)
ICA	Colombian Agricultural Institute (Colombia)
IDB	Inter-American Development Bank (Latin America)
IDEAGRO	Agro-industrial plant of the IDEAS Centre (Peru)
IDEAS	IDEAS Centre (Peru)
IDRC	International Development Research Center (Canada)
IER	Institute for Rural Education (Chile)
IFAD	International Fund for Agricultural Development (Italy)
IFC	Research and Training Institute for Cooperatives (Honduras)
IHDER	Honduran Institute for Rural Development (Honduras)
IICA	Inter-American Institute for Agricultural Cooperation (Latin America-Costa Rica)
ILDIS	Latin American Institute for Social Research (Latin America)
INC	National Institute for Colonization (Bolivia)
INDAP	Institute for Agricultural Development (Chile)
INIA	Institute for Agricultural Research (Chile)
INIAA	National Institute for Agricultural and Agroindustrial Research (Peru)
INIAP	National Institute for Agricultural Research (Ecuador)
INPROA	Institute for Agricultural Development (Chile)
IPDS	Private institutions of social development (Bolivia)
IRDP	Integrated rural development programme
IRENA	Natural Resource Institute (Nicaragua)
ISNAR	International Service for National Agricultural Research
IU	Intermediate user
MACA	Ministry for Campesino Affairs (Bolivia)
MAG	Ministry of Agriculture and Livestock (Ecuador)

MAG	Ministry of Agriculture and Livestock (Nicaragua)
MIRENEM	Ministry of Natural Resources, Energy and Mines (Costa Rica)
MSO	Membership service organization
NARS	National agricultural research and extension service
NGO	Non-governmental organization
ODA	Overseas Development Administration (UK)
ODI	Overseas Development Institute (UK)
OFCOR	On-farm client-oriented research
OFPR	On-farm participatory research
OTS	Organization of Tropical Studies
PHS	Protected Horticultural Systems
PIP	Programme for Production Research (Ecuador)
PROCADE	Programme for Alternative Campesino Development (Bolivia)
PRODESSA	Agricultural Centre for Research, Development and Training (Nicaragua)
PRONADER	National Programme for Rural Development (Ecuador)
PRN	National Reconstruction Plan (El Salvador)
SEPAS	Diocesan Secretariat for Pastoral and Social Work (Colombia)
SIPA	Agricultural Research and Development Service (Peru)
TAU	Technical advisory unit
UNCED	United Nations Conference on Environment and Development
UNITAS	Union of Institutions of Social Work and Action (Bolivia)
USAID	United States Agency for International Development (USA)

1

INTRODUCTION

A NEW VISION OF RURAL DEVELOPMENT? ENTER THE NGOs

The recent explosion of interest in non-governmental organizations, or NGOs, seems set to continue, as they become the new darlings of donor agencies. Why is this happening? It is certainly not because they are a new phenomenon in Latin America. In the 1960s and 1970s indigenous non-governmental organizations were already making important contributions to political resistance, social welfare and grassroots action. But it was only in the late 1980s that they came of age, and into vogue. Donors, faced with the crises and inefficiencies of their traditional governmental counterparts, now want to work with NGOs in programmes of poverty alleviation, good government and sustainable development. Social scientists, faced with the crisis of their former models of social development, have latched on to the work of NGOs for theoretical inspiration and political hope. In the last few years NGOs have been denominated vehicles of development, democracy and empowerment at the grassroots.

The interest in NGOs' potential contribution to rural and agricultural development stems from the growing concern, in all circles, for new strategies of agricultural development. Orthodox models seem to lie in tatters. A new technological agenda is urgently required to meet the combined challenges of sustainability and the increasing competitive pressures on agriculture unleashed by trade liberalization and regional integration. New institutional configurations are needed because acting alone, the Latin American state, structurally adjusted and reduced in size, will simply be unable to meet contemporary challenges in agricultural development – particularly if a more decentralized institutional approach is required to respond more efficiently to local needs. Finally, if agricultural development programmes are to contribute to the challenge of good government, then they too will have to make themselves more accountable and participatory.

These new visions have generated interest in NGOs. This is partly because the challenges posed play to strengths which many have argued are characteristic of NGOs. NGOs have, many claim, been at the forefront of efforts to develop more appropriate, lower-cost and agroecologically-sound technologies. Perhaps more importantly, NGOs appear to have particular strengths in working with the rural poor, favouring a participatory approach to rural development in general and agricultural technology development (ATD) in particular.[1] Many of them espouse a concept of participation that elicits farmer ideas, experiments and evaluations, and link concepts of participation to wider concerns for strengthening the rural poor's capacity to articulate their demands to the state and to press for broader democratization of local development administration. Such an approach to participation would, if realized in practice, make positive contributions to technology development and local governance at the same time.

The interest in NGOs is also inspired by more instrumentalist concerns. Many donors and governments seem to perceive NGOs as a means of filling gaps in public programmes opened up as the state withdraws from different development activities. In these more instrumentalist visions, NGOs will be vehicles for implementing programmes previously executed by the state.

It has not simply been that the state has shown more interest in NGOs. Many among them and their sympathizers have likewise begun to show more interest in the state. As people have begun to look more closely at NGOs, they have not only asked 'What do NGOs do well?' They have also begun to ask 'How might they do more of it?'; or, to use another phrase in fashion, how might they 'scale up' their impact? Answers to these questions have taken many back to the public sector. On the one hand, they see it as a source of support to strengthen NGO work: access to the technologies of government research stations could for instance improve the quality of NGOs' own agricultural development work. More critically, many stress that if NGOs are to have any significant impact then they must ultimately seek to change the administrative structure of the state, and then aim to work with it so as to disseminate their ideas and proposals more widely.

Commentators on agricultural and rural development in Latin America have therefore argued the need for closer coordination between the public sector and NGOs in order to capitalize on the comparative advantages of both. Thus, claims Osvaldo Barsky, Argentine economist and rural sociologist, 'It is a matter of priority that Latin American governments take advantage of this network [of NGOs] which already exists and which has considerable accumulated experience' (Barsky 1990: 74). Barsky's observation is echoed by Fausto Jordan, an Ecuadorian whose career has taken him through executive positions in NGOs, ministries, donor agencies and international agencies: 'The existing structures and work methods of the NGOs can be utilized with good effect to complement the efforts of government bodies in rural development actions' (Jordan *et al.* 1989: 269).[2] These different propositions seem to be moving us towards a vision of rural development in the 1990s with a state reduced in size, supporting and coordinating the actions of a gamut of NGOs and local popular organizations in a joint development effort.[3]

A CRITICAL REFLECTION

These visions are attractive – but they have various shortcomings. The most obvious limitation is that they remain very general and do not delineate the mechanisms by which NGOs would collaborate with, or relate to, government. Yet clearly such mechanisms would have to be tightly defined, their potential drawbacks understood and political obstacles to their success made explicit. A second limitation is that they rarely discuss the trade-offs that might be involved for NGOs if they collaborate with government. What might they lose in such relationships? What might collaboration imply for institutional identity and autonomy? If, as is the case in Latin America, many NGOs have spent years resisting authoritarian governments, what will it mean for them to make an about turn and begin working with governments? The change, we might imagine, is likely to be complicated and painful.

These issues have to be addressed before we can talk glibly about the virtues of NGOs working with governments. Furthermore, many commentators tend to make generalizations about the relationship between NGOs, the state and the rural poor, that do not recognize the diversity that exists in the NGO community. Yet if NGOs are diverse, it can be assumed that this would allow, and indeed require, different forms of state–NGO relations. Moreover, claims about NGOs' supposed qualities tend to be accepted at face value. In order to develop proposals for inter-institutional coordination that will make any impact on rural poverty, such claims merit a more critical analysis.

Finally, some of these viewpoints often give the impression that forging more collaborative NGO–public-sector relations will be a relatively simple process. They skate over the problem of politics and seem to assume that NGOs, government and grassroots organizations will have like minds on social and development policy. Yet the history of tense and conflictive NGO–state relationships suggests otherwise, as does a whole series of contradictions that have to be faced by NGOs as they consider working with governments. Consider just one example of these contradictions. Latin American NGOs have argued since their very beginnings that real democracy demands that the state refocus its efforts on marginalized sectors and assume social responsibility for their welfare. Given this perspective, one might expect it to be difficult for NGOs to accept a concept of collaboration with the state wherein they themselves relieve it of this social responsibility and implement agricultural and rural development programmes on its behalf.

THE STUDY'S GOALS

Out of such sympathetic but critical reflections on the new visions of rural development came the study which we will describe in this book. Its main purpose was to develop the thinking behind these new visions, analysing the contextual factors present in wider society and the agricultural sector which impinged both on the possibility of a more coordinated NGO–public-sector relationship, and on the contribution of such a relationship to poverty alleviation and to the development of the peasant *campesino* economy (we use the Spanish word 'campesino' to refer to poor rural people who derive a significant part of their food or income from agricultural activities). The study was based on several broad postulates. First, we felt that the NGO–state relationship in ATD would be influenced by many factors not directly related to agricultural technology, and so would be far more complex than many proposals for collaboration have suggested. We believed it would involve trade-offs for NGOs which need to be made explicit. However, we also postulated that within certain contexts, a stronger, coordinated relationship between public-sector and NGO efforts is feasible and has the potential to influence research, extension and a wide range of social factors impinging on the campesino sector so as to make the system for ATD in Latin America much more effective and relevant for the rural poor than it has been in the past. Benefits could be derived from exploiting the different comparative capacities of NGOs and public-sector agencies in agricultural development (such as the national research and extension services, or NARS) by way of linkage mechanisms. However, we believed these mechanisms must be set out in more detail than has been the case to date if such benefits are to be realized. The third postulate was that involving NGOs in public-sector programmes had the potential to orient these programmes more strongly towards the needs of campesinos, reversing some of the current focus and biases of the NARS programmes towards wealthier farmers and modern technologies. It might also enhance NGO effectiveness.

Such changes will not come easily. One of our prescriptions has been that in order to achieve closer NGO–public-sector relations after so many years of mutual suspicion, requires a change in attitudes on the part of personnel in both public institutions and NGOs. Consequently, we also tried to make the study something more than a simple research project, and to an extent the research became engaged in the very process of promoting change in inter-institutional relationships. Through the study, we sought to facilitate relations between NGOs and public-sector institutions, promoting flexibility in attitudes that are sometimes highly critical and dogmatic, and trying to foster an improvement in NGOs' and governments' mutual understanding of the comparative advantages, the limitations and the nature of the activities undertaken in both sectors. We also hoped to help improve donors' understanding of the nature of the NGOs, and to contribute to new proposals for the inter-institutional organization of ATD in Latin America, based on this public sector-NGO–relationship.

METHODOLOGY OF THE STUDY

The main part of the study was a strategic selection of case studies in five countries: Bolivia, Chile, Colombia, Ecuador and Peru. The case studies were selected on the basis of discussions with key people in the region, and the personal knowledge of the authors. The prime criterion for choice was that the cases should highlight experiences that were significant for the themes of the study and should offer valuable lessons either about public-sector—NGO relations or about the nature of NGOs and their work in ATD. The information gained from these studies was complemented by a series of interviews with a wide range of public, donor, non-governmental and campesino institutions which have an interest in ATD in the different countries. These interviews, along with literature reviews, allowed us to explore further the general context of NGOs' agricultural work and their relationships with government. Three types of case study were chosen:

- *Public sector.* Cases selected from the public sector were restricted to the national institutions responsible for the development and transfer of technology, the NARS. The study also recognizes the importance of rural development projects and other public-sector actions in rural areas, and these are discussed in the text. However, our focus on ATD led us to concentrate on the NARS.

  Whilst a considerable number of studies on ATD in the public sector have been carried out,[4] there has been much less research on ATD in the NGO sector. Therefore we chose only those NARS cases which could throw light on the public-sector–NGO relationship, or on attempts by the NARS to institutionalize participatory research, and the implications of such attempts for their relationships with NGOs.
- *NGOs.* The case studies are of NGOs which have an important role in the development and transfer of technology for campesino agriculture, even if this is rarely the sole focus of their work. However, the range of organizations which are termed NGOs is broad and so our use of the label merits a comment (Fig. 1).

  In literal terms, 'non-governmental' could include the commercial private sector, although the label 'NGO' is rarely applied to businesses. The importance of this commercial sector is growing in Latin American ATD. Much biotechnology research is done by private business, as is much seed production. Similarly the private sector is the source of agrochemical pesticides and many fertilizers. At a local level it is represented by input distributors. While these clearly have a role in agricultural development, their motives are narrowly defined and certainly do not embrace questions of local participation and campesino welfare and organization. Furthermore, their emphasis on selling agrochemicals does not predispose them to promote low-input practices, or the more efficient input use that is crucial for economic and environmental reasons. Finally, the complex and diverse

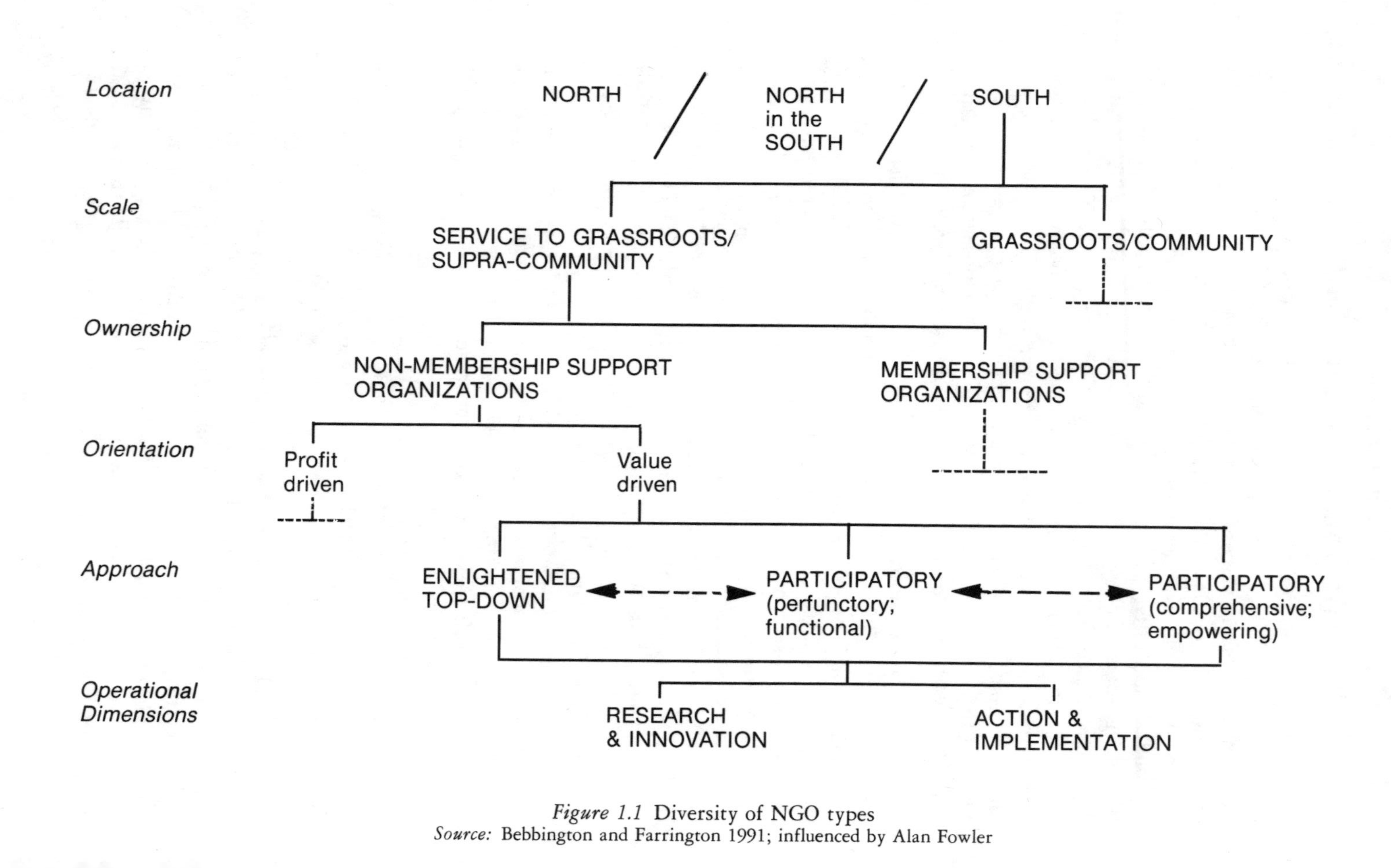

Figure 1.1 Diversity of NGO types
Source: Bebbington and Farrington 1991; influenced by Alan Fowler

agroecological and socioeconomic conditions in which the rural poor live mean that these are highly fragmented markets, often distant from commercial centres and costly to reach. Overall the private commercial sector is in the business of profit-making, not poverty alleviation, and so will be only marginally interested in these areas. It is therefore not considered in this study.

The term 'NGO' is conventionally used to describe both national organizations and international organizations (such as CARE or OXFAM) based either in the North or South. In this study we concentrate on national organizations, since their consolidation as institutions is a vital ingredient for the strengthening of Latin American civil society and a more effective local development process. Among these national organizations the term is still applied to a wide variety of really rather different types of institution. A first distinction can be drawn between those that are grassroots organizations, and those that are grassroots service organizations. The former type of organization, such as communities, cooperatives and neighbourhood associations, are actors meeting their own needs. These 'base' groups differ from those NGOs which give services to a number of different base groups. These types of NGO act primarily to meet others' needs, not their own. Following Carroll's terminology (Carroll 1992), such service or support organizations (what he collectively calls 'intermediary NGOs') can be broken down into 'membership' and 'non-membership' categories.

Membership support organizations (MSOs) are staffed by individuals elected from and by the grassroots organizations that are themselves members of these supra-communal MSOs. Non-membership support organizations, or what Carroll calls 'grassroots support organizations' (GSOs) are not elected, and are by and large staffed by urban professionals who are often socially and ethnically distinct from the grassroots. Consequently the management styles and sociology of MSOs and GSOs are quite distinct, as are the dynamics of their relationship with the grassroots.

In this study only one MSO is included, El Ceibo of Bolivia. Instead, our analysis focused on GSOs, for reasons elaborated in Chapter 2. When we use the term NGO, we are then speaking of these types of support organizations.[5]

The selection of case studies was further limited to progressive (or value driven) NGOs, many of which had been set up during a period of dictatorship or political repression. These organizations emphasize the objective of working alongside and together with the campesino movement. These are the organizations which have referred to themselves as 'development NGOs' (Arbab 1988); 'private institutions for social development' in Bolivia (Velasco and Barrios 1990); and 'private development centres or associations' in Peru (Padrón 1982). Cases were selected from those NGOs which demonstrated considerable institutional solidity, and which had a presence in many regions of their countries and were active and significant

Table 1.1 The case studies[7]

	Public sector	*NGO*	*NGO coordinating body*
Bolivia	CIAT IBTA	CESA CIPCA (Kohl)	PROCADE
Chile		AGRARIA GIA	
Colombia	ICA	FUNDAEC SEPAS	
Ecuador	INIAP-PIP	CAAP CESA	
Peru		CIED DEAS	CCTA
Latin America			CLADES

Note: See Annex 1 for full names.

at a national level. We chose two broad types of case: those where there had been a significant experience of the relationship between the NGO and the NARS and other public institutions (AGRARIA, CESA-Ecuador, CIED, CIPCA, FUNDAEC, GIA, IDEAS, SEPAS) and, second, those cases where the experience of institutional isolation and competition had had important implications for the ATD work of the NGO or for coordination among NGOs (CAAP, CESA-Bolivia, the Kohl study). The full names of these NGOs and case study titles are given in Annex 1.

In this sense, our sample was not representative – there exists a multitude of NGOs which are much weaker, smaller and less progressive than the NGOs analysed here. Because of this, some care is needed in interpreting and generalizing from our findings. Nevertheless, the case studies still give important lessons for the general analysis of NGO–public sector-relations.[6]

- *NGO coordinating bodies*. Finally, we selected some cases of NGO coordinating bodies, institutions that have become increasingly important in the last decade. These are organizations which facilitate communication between different NGOs and may also seek to link their actions. The studies covered two types: organizations focusing on the NGO context within a country (CCTA, PROCADE) and those looking at Latin America as a whole (CLADES). In addition, they highlighted some implications – both positive and negative – of seeking more formal coordination among NGOs. This, in turn, had implications for NGO relations with both funding agencies and with the public sector.

The organizations considered in the three types of case study are listed in Table 1.1.

The case study authors were asked to address a range of issues. First, they discussed the sociopolitical origins of the institution analysed, and its place within wider sociopolitical and economic change in its country. They then considered the evolution of the institution's relations and programmes with campesinos, and the specific role that agricultural technology development had played in the programmes. They described the nature of those ATD activities and the particular problems that had arisen. It was in these contexts that inter-institutional relationships (NGO–state, NGO–NGO, etc.) were considered. Were they politically or technically motivated? Were they collaborative, conflictive or (as was most usual) a bit of both? What form did they take? What were the impacts, good and bad? What does the institution now feel about the role of such relationships for its future work? This material was the basis for the interpretations offered in this book.

We have not reproduced case studies in full in the book, and have chosen instead to draw on them selectively. Summary profiles of eight case studies are, however, presented in Annex 2.

The studies were written or supervised by NGO or government staff (often by a member of the executive), who were in permanent contact with one of the study's co-ordinators.[8] This allowed a dialogue during the preparation of the study. This method had several advantages. First, rather than sending in outside authors or consultants, NGOs were given their own voice in the research. Second, it meant that case studies were built on the personal knowledge of key individuals within the organization. Conversely, of course, their biases and personal interpretations were also incorporated. One of the key roles of the coordinators was therefore to be a 'devil's advocate' and a more independent commentator – both in the preparation of the case material and in the interpretation and analysis in this book.[9]

The material was subsequently discussed in December 1991 at a workshop held in Santa Cruz in Bolivia, involving the institutions involved in the study and a range of other people from NGOs, government and donor agencies (Bebbington *et al.* 1992b). This workshop brought together people who have often been highly critical of each other's institutions, and some of that tension was undoubtedly played out in Santa Cruz. On the other hand, the discussions were remarkable for the degree of self-criticism that NGOs and public-sector staff were willing to make. This self-criticism was also the basis of a largely shared conclusion that some form of closer interaction between all involved in agricultural development would be desirable. Nobody, however, was under any false illusion that such change would be simple. The discussions at this workshop, and the dialogues during the course of the year, have greatly influenced the preparation of this book.

ORGANIZATION OF THE BOOK

The chapters that follow present the main themes addressed in the case studies, but they are organized so as to argue a position based on our interpretation of NGO–state relations as revealed by this and other material collected during the research. Following this introduction, Chapter 2 lays out in more detail the main concepts on which the book is based. In schematic terms it shows how we have attempted to draw on two bodies of writing on Latin America that are rarely brought together: one on state–society relations and another on the design of agricultural development.

Chapter 3 then focuses on the process through which new visions of the institutional organization of agricultural development have emerged. It analyzes how sociopolitical and economic processes have affected the emergence and the nature of NGOs and NARS in the countries studied, as well as their wider objectives and the interrelationships between them. The chapter analyses the role of these actors in ATD in the light of the more general roles of the state and NGOs in political change. Until recently these roles have cast NGOs and state in opposition to each other: the state has repressed civil liberties, and NGOs have acted as vehicles of political resistance and criticism. However, we suggest that recent changes in macroeconomic and political context have called such former conflictive relations into question. They are beginning to force reflection on the issue of NGO–state collaboration.

Chapter 4 then analyzes the technical changes in Latin American agricul ure that have generated criticisms of orthodox Green Revolution models, and led to the elaboration of a new technological agenda for agricultural development. At the same time as the relations between NGOs and the state were changing, Latin American agriculture was going through a rapid process of agricultural modernization. We argue that, due to a variety of socioeconomic and agroecological factors, this process of technological and social change has not fully addressed the technical and social needs of campesinos, even if they have not been entirely excluded from its benefits. We outline two perspectives on how this modernization needs to be reoriented in order to generate and aid the adoption of technologies more appropriate to campesino needs. The two perspectives, one proposing a scaled down and more inclusive modernization, and the other a more radical agroecological break with the past, have differing approaches to technology. However, we suggest that they share much in common as regards their proposals for ATD strategies and the role NGOs should play in these. Their overlap is the basis of an emerging technological agenda for the 1990s.

As actors in this process of technical change, NGOs and NARS have sought to work with, and develop technological agenda for, the rural poor. In the fifth chapter the focus turns to their more strictly agricultural development activities and assesses how they have in fact performed in the past, and whether the two proposals discussed in Chapter 4 reflect fair representations of what NARS and NGOs have done and could contribute in ATD. The differences between the technical agendas of NARS and NGOs, their capacities to carry out the different activities in ATD, and the institutional characteristics which affect these capacities are outlined. This chapter concludes that the NGO and NARS sectors do, with certain reservations, reflect the qualities that the new visions of agricultural and rural development have apportioned to them. Each demonstrates particular strengths and weaknesses, which could complement each other. It is argued that a change in inter-institutional relations to overcome the limitations of the respective actors and to foster linkages between them, could improve the nature and availability of technology for campesino development. It could also enhance the ability of the campesino to make full use of such technology. However, the prior and contemporary sociopolitical contexts outlined in Chapter 3 will mean that taking advantage of such functional specialization will not be straightforward.

Chapter 6 begins by examining some past attempts by NGOs and NARS to respond to the limitations on their abilities to contribute to the generation of appropriate technologies. Initially their responses involved institutional changes specific to each: such as the incorporation of on-farm research and farming systems programmes in NARS, and the development of coordination mechanisms among NGOs. NGOs also sought at times to scale up their impact by exercising pressure over NARS through certain advocacy activities. Direct and sustained contacts between NGOs and NARS were, however, rare until recent years. Now, though, we are seeing more fundamental changes in inter-institutional relationships, in large part as a result of major political and macroeconomic changes and pressure from donors. The most significant changes have been the economic crisis; the reform of the public sector under the exigencies of structural adjustment; the return to democratic rule; and the desire of donors to strengthen the state and civil society in different spheres. The changes, and particularly the quite radical restructuring and weakening of the NARS, are forcing the issue of NGO–NARS links in ATD, and opening possibilities for what could be a more efficient relationship and a more democratic process of agricultural development.

Any success in taking advantage of these opportunities will depend on how effectively and skilfully different inter-institutional linkage mechanisms are installed. Chapter 7 reviews the available evidence on the different forms of NGO–government interaction that have been experienced to date, and considers their effect on the capacity of all actors in ATD to generate appropriate technologies. The chapter concludes with a discussion of the key areas which NGO and NARS directors would need to consider before committing themselves to such linkages, and selecting the most appropriate mechanisms for them.

Chapter 8, an invited chapter written by David Kaimowitz of IICA, shifts our geographical attention to Central America and gives an overview of these same aspects of NGO and NARS activity in that region.

Chapter 9 draws conclusions about the key characteristics of NGOs and NARS in the process of ATD in Latin America. The chapter offers some reflections on the changing roles of NGOs and NARS in Latin American rural development, and how these might be understood in broader terms. It then makes concluding considerations as regards the positive, negative and ambiguous implications of a link between the different actors, particularly as regards their implications for agricultural development, rural democracy and institutional identities. On the basis of this summarized assessment, the chapter offers our interpretations of, and recommendations for, future paths of ATD and its inter-institutional organization. These recommendations are aimed at the heads of NGOs and NARS, and also at the funding agencies who argue for closer collaboration between the NGOs and the NARS.

NOTES

1 We use the term 'agricultural technology development' to refer to the process in which agricultural technology is generated, tested, disseminated, adopted and continually adapted by farmers, professional researchers, and extensionists.

2 All quotations originally in Spanish have been translated by the authors.

3 'The state should manage a macro-context and a social policy for rural development which involves the balanced delivery of public goods and services . . . management of projects must be left to NGOs and grass-roots organizations, in line with their comparative advantages . . . their initiatives should be coordinated both amongst themselves and with public programmes' (de Janvry *et al.* 1989).

4 Most notable among these are the many studies and consultancies conducted by ISNAR, the International Service for National Agricultural Research.

5 Several of the authors had also been involved in evaluations of the resource management work of MSOs (e.g. Bebbington *et al.* 1992a), and these studies served as a counterpoint to some of our material on GSOs.

6 There exist other studies which offer such lessons for the weaker NGOs (e.g. Kaimowitz *et al.* 1992).

7 See the List of Abbreviations for the full names of NGOs. The Kohl study is an analysis of ATD work among a range of NGOs, and seeks to draw certain conclusions regarding their strengths and weaknesses.

8 These were Anthony Bebbington, Penelope Davies, Martin Prager and Graham Thiele.

9 This book draws selectively on these studies. Some have been published in full in English (Aguirre and Namdar 1992; Kohl 1991; Sotomayor 1991). The complete set of studies in Spanish is held at ODI in London and CIAT in Santa Cruz (addresses given in Annex 1). Other studies relevant to the theme and developed with ODI are Bebbington 1989, 1991; Berdegue 1990; Thiele *et al.* 1988.

2

DEMOCRATIZING AGRICULTURAL DEVELOPMENT?

Concepts for analysing NGOs, the state and agricultural development in Latin America

RURAL DEMOCRATIZATION AND RURAL DEVELOPMENT

'Most research systems in developing countries were organised to serve commercial farmers operating in more favourable and homogeneous agroecological conditions than those in resource-poor farming contexts' (Merrill-Sands and Kaimowitz 1990: 1). This observation comes from the coordinators of what have been perhaps the two largest studies on the organization of agricultural research and extension in the public sector.[1] The implication is that research systems as currently organized are not likely to generate technologies appropriate to the needs and circumstances of the rural poor farming in so-called complex, diverse and risk-prone environments (Chambers *et al.* 1989).

This situation is undesirable on equity grounds. It also implies that research systems are, as traditionally structured, ill equipped to respond to the principal pillars of contemporary development programmes and aid policy: democratization, poverty alleviation, sustainable development and economic growth. Let us briefly consider each in turn.

Many of the discussions of *democratization* and state–society relations suffer from excessive levels of abstraction. If we ground them more specifically in ATD, what might be meant by the democratization of agricultural development? It obviously means that the bias of research systems to commercial farmers is unacceptable. Nevertheless, it ought to signify more than equity in the distribution of the fruits of state intervention in the countryside (see Fox 1990a). It should imply the creation of means through which campesino interests can be represented in the institutions responsible for agricultural and rural development in order to make them more responsive and accountable to the rural poor. It ought also to imply the need for more direct forms of campesino participation in actions oriented towards agricultural development. In Bobbio's terms, it would require an enhancement of both 'representative' and enhanced 'direct' democracy: the former strengthening campesino representation in official decision-making, the latter 'increasing the number of contexts outside politics where the right to vote is exercised' (Bobbio 1987: 56).

Identifying mechanisms for making public institutions more accountable, transparent and subject to the demands of broad-based participation in the political process, is justified not only on the grounds that development loses much of its social meaning if it is not accompanied by democracy, enhanced equity and popular empowerment (CEPAL 1990; Lehmann 1990; Friedman 1992). It is also justified by more pragmatic concerns. The tasks of economic reactivation and growth are that much more difficult without more democratic and inclusive forms of society. If campesinos are not allowed to participate more fully in society, there is more possibility of social protest that will upset any growth that current neo-liberal models might otherwise stimulate (de Janvry *et al.* 1989).[2]

It seems similarly reasonable to argue that if '*poverty alleviation*' is a goal of development programmes, then ATD programmes in rural areas – which is where poverty is still concentrated (Table 2.1) – must be far more inclusive than they have been to date (see CEPAL 1990). Given that agricultural activities provide the basis to the livelihoods of many of the rural poor, then poverty alleviation programmes must include technological, infrastructural, socioeconomic and other interventions aimed at enhancing the security and productivity of food and cash-crop production. An ever-growing body of information suggests that such interventions are more successful when local groups have more say in their design and administration (Chambers *et al.* 1989; Oakley 1991). Furthermore, the growing debates on *sustainable natural resource management* and environment and development have demonstrated that local groups are often the most knowledgable and committed stewards of local resources (Goodman and Redclift 1991; World Bank 1992).[3] Consequently the goal of sustainable development will also be furthered when local groups have more influence over the development of technologies for 'fragile lands' (Browder 1989) and over the way the environment is used – although frequently other structural imbalances in land distribution and pricing policies also need to be addressed to offset environmental degradation in campesino areas.

Finally, perhaps the most convincing argument for more inclusive rural development is that the campesino sector remains *economically* important, particularly as a producer of basic foodstuffs for domestic markets (CEPAL 1982; Grupo Esquel 1989).[4] According to the Andean Pact, campesino agriculture in the Andean area generates between 50 per cent and 60 per cent of all agricultural consumption goods (CEPAL/FAO 1986). Nor should the importance of campesinos in the production of certain export crops, such as coffee and cacao, be underestimated: in Ecuador, 65 per cent of cacao production comes from campesino landholdings, and in Peru, 54.8 per cent of total coffee production (Jordan *et al.* 1989).[5]

Table 2.1 Population, production and poverty in Latin America

	Rural population (millions)[a]	*Percentage of population in rural areas*[a]	*Percentage economically active population in agriculture*[b] *(1990 figures)*	*Percentage of rural population in poverty*[c,d]	*Campesino production as per cent of gross value of agricultural production*[e]
South America					
Bolivia	3.51	49.3	41.2	85	80
Chile	1.92	14.8	12.5	20	38
Ecuador	4.54	44.0	30.3	65	–
Colombia	9.80	30.3	27.6	70	44*
Peru	6.43	30.4	34.7	83	55
Central America					
Costa Rica	1.52	55.5	23.9	20	–
El Salvador	2.87	55.9	37.4	32	–
Honduras	2.85	57.2	55	70	–

[a] World Bank 1990b.
[b] World Demographic Estimate and Projections 1950–2028, UN 1988 (1990 figures are projections).
[c] UNICEF 1992.
[d] UNICEF 1990.
[e] FAO/CEPAL, 1986, and CEPAL, 1985 cited in Jordan *et al.* 1989, p. 225.

On the basis of the concentration of productive potential and poverty in the campesino sector, some have argued the need to identify a *via campesina*, or 'campesino path', to agricultural development in Latin America – a development based on campesino production units (Figueroa 1990). Theoretically, this argument builds on a tradition of Chayanovian analysis that stresses the rationality and efficiency of peasant production units, and above all, their resilience in surviving and continuing to produce in circumstances which capitalist enterprises would deem unprofitable. Strategically, such an option would constitute a recognition of the campesinos' economic importance and a response to poverty in the zones predominantly inhabited by campesinos (Jordan *et al.* 1989). One of its pillars would have to be the development of technologies that can be incorporated into campesino production units and which would increase their efficiency and competitiveness in the market.

RESPONDING TO THE CHALLENGES: NGOs, INSTITUTIONAL CHANGE AND A NEW TECHNOLOGICAL AGENDA

There are, then, many good reasons for increasing the orientation of NARS and dominant ATD institutions to the rural poor. Nevertheless, calls for change will not, by themselves, get us very far. The institutional conditions for what we labelled a 'new technological agenda' (Kaimowitz 1991) will not spirit themselves out of thin air, nor out of good will alone. Nor will appropriate technologies alone solve rural poverty. Mechanisms must be found for generating these technologies, for ensuring that their continuous generation is institutionalized, and for combining technological interventions with other elements of socioeconomic change.

It is in the search for these mechanisms that NGOs are attracting attention. There are perhaps two main contexts in which they may contribute to these mechanisms – first, through their relationships with the NARS and, second, through their actions in rural civil society. They many contribute to these mechanisms through their relationships with the NARS in two ways:

- by exerting pressure over NARS so that they reorient their work to campesinos, and incorporate methods, institutional arrangements and technologies that NGOs have developed;
- by entering working relations with NARS in order (1) to increase the appropriateness of the work of the NARS or (2) to enhance the effectiveness of the NGOs' own work.

They may contribute through their actions in rural civil society:

- by acting independently as development institutions responsive to campesino needs,
- by acting independently as advocates for changes in policies that have frequently been detrimental to the rural poor (de Janvry and Garcia 1992);
- or by, in Thomas Carroll's terms, 'building capacity' at the grassroots (Carroll 1992).

'Capacity building' is a broad concept in Carroll's hands, but revolves around two main principles. The first is to enhance campesino self-management abilities. In the case of ATD this would include the creation of local organizations which could build on, improve and then disseminate the results of campesino experiments to other rural people. The second is to increase the poor's political capacity to hold institutions to account and to press for social change. For Carroll, these are at once the essence of, and essential for, grassroots development; in the particular context of this book, they would seem essential prerequisites to achieving, and ensuring the beneficial impact of a more equitable, more democratized ATD. The main conclusions of Carroll's 1992 study is that NGOs' *main* contribution and strength is that they build this capacity.

It is useful to elaborate how some of the most relevant discussions in the literature have addressed NGO contributions in these two broad contexts. While the different roles of improving NARS' capacity to generate appropriate technology, of acting as advocates for policy change and of building grassroots capacity are not separate, we will treat the themes separately in the following sections. This is partly for ease of presentation and partly because discussions of the two have themselves been conducted somewhat separately.

DEVELOPING APPROPRIATE AGRICULTURAL TECHNOLOGIES: NARS AND NGOs

The largest and most comprehensive studies on the organization of research and extension for campesino agriculture have been two research programmes coordinated by the International Service for National Agricultural Research (ISNAR) since the mid-1980s (Merrill-Sands and Kaimowitz 1990; Kaimowitz 1990). While not restricted to Latin America,[6] the studies generated a number of arguments and concepts relevant to this book, and we have drawn on the ISNAR conceptual framework to organize our own material, in order to facilitate comparison between the different studies. We return to our own elaboration of this conceptual framework in the final section of this chapter.

The ISNAR studies were based on two main premises. The first was that the ATD system was much more likely to generate and transfer agricultural technologies appropriate to small farmers if the system incorporated an element of on-farm client-oriented research (OFCOR) (Merrill-Sands 1989). Although there could be different degrees of farmer participation in such OFCOR, ranging from the simple execution of trials on farmers' fields, through to research processes in which the farmer–researcher relationship is conducted as if between equals (Biggs 1989), the general principle was that participation would increase the appropriateness of technologies. The second premise, the subject of a different study (Kaimowitz 1990), was in some respects an elaboration of the first. This was that the generation of appropriate technology is more likely where the links between researchers, technology transfer agents and farmers are strong and allow two-way flows of information.

The studies were devoted primarily to understanding how such OFCOR and linkages could be strengthened and institutionalized within and among different parts of the NARS. Enhancing participation and strengthening linkages were treated mainly as management issues (e.g. asking questions such as how research managers could improve the research–extension link). None the less, it was also recognized that the degree and quality of linkage and of farmer participation also depended on contextual factors that were not directly within the ambit of research managers' normal spheres of influence. We feel such contextual factors merit much more attention in explanation of ATD institutions in Latin America.

Three types of contextual factor were recognized by ISNAR: external pressure on NARS, resource availability and agroecological diversity. Regarding the first of these, the study coordinators comment, 'external pressure is not the only thing which motivates institutions to develop and deliver relevant technologies to resource-poor farmers, but it is an essential ingredient in the process' (Merrill-Sands and Kaimowitz 1990: 21). While it was argued that campesinos themselves have rarely exercised such pressure (Sims and Leonard 1990), it was also suggested that others, such as NGOs, may do so on their behalf (Merrill-Sands and Kaimowitz 1990: 24). Resource availability, the second contextual factor, was deemed critical in linking research, extensionists and farmers because, quite simply, such linkages cost time, money and effort. Where resources are scarce, linkages in NARS are likely to be weakened. One suggested response to this constraint was to 'use' NGOs and other private organizations to carry out technology testing and dissemination, thus freeing up resources in the public sector for linkage mechanisms. Third, the particularly diverse agroecological environments that characterize many developing countries mean that it is difficult for NARS to address all local needs. It is simply impossible for the NARS to cover the cost of generating technological recommendations for all these environments. In such circumstances, it is implied, more location specific institutions (such as NGOs) would be better able to respond to local needs. Thus, despite their public-sector focus, the ISNAR studies recognized that NGOs could have important roles to play in conjunction with the NARS in generating and transferring appropriate technology in contexts of agroecological diversity, public-resource scarcity and where campesino groups are unable to exercise much pressure over the NARS – situations that are all encountered in the Latin American countries of this study. ISNAR, then, began to pose the question of links between NGO and NARS in ATD – it did not address how such links might be put into operation.

Another finding of ISNAR's and other studies is that NARS have had difficulty in furthering and institutionalizing participatory approaches to technology development. A number of authors have suggested that NGOs are far more capable, and willing, to promote grassroots participation in their activities (Farrington and Biggs 1991). These suggestions have been a further source of interest in NGOs. Linking NGOs into government programmes might help make them more 'participatory' (World Bank 1991a).

However, as noted in the Introduction, many NGOs have a more radical understanding of participation than government agencies might be willing to accept. It is one thing for a NARS to want participation to help generate adoptable technologies; they are far less likely to want to introduce participatory approaches that will imply more accountability to the rural poor. Linkages with NGOs to promote 'participation' may therefore generate tensions, indeed conflicts, when it becomes clear that NARS and NGOs mean quite different things by the word 'participatory'.

This potential source of tension is related to a more general difference of emphasis in the interest in NGOs. Some imply that NGO involvement in NARS programmes will be the missing ingredient in the recipe for successful ATD, adding participation, local knowledge, money, and pressure for change. They assume the ingredients will mix well, but there is no reason why they should. While government and donors might like NGOs to add resources to help NARS work, NGOs may feel the state should pay for that work.

Furthermore, some talk glibly of linkage mechanisms for enabling both collaborative activities (e.g. dividing tasks in research and extension between NGOs and NARS) and more potentially conflictive activities (e.g. NGOs pressuring the NARS to change the focus of their work). While it may be possible to manage both collaboration and conflict through the same relationship, there is no reason to assume that this may always be possible. It is easy to imagine how an NGO's attempt to exercise pressure on a NARS could undermine collaborative relationships.

Aside from this caveat, there are other reasons why proposals for NGO–government linkages should be treated with some caution. For it is also the case that the empirical support for claims that NGOs have valuable lessons for wider-scale implementation, that they have the capacity to generate and transfer technology, and are skilled in fostering popular participation, remains weak. Many of these claims derive more from rhetoric and received wisdom than from evidence. Similarly, the organization of these proposed NGO–NARS links remains uncertain, as does the willingness of NGOs to assume them.

NGOs IN RURAL CIVIL SOCIETY: BUILDING GRASSROOTS CAPACITY

The changes that must occur in rural Latin America in order for poverty to be alleviated, development to be more inclusive and local democracy to be more solid, go well beyond the generation of appropriate technology. These wider changes are not the specific theme of this study, but they must be discussed because they themselves increase the likelihood of ATD being more appropriate and beneficial for campesinos. They are also crucial if a 'participatory' ATD is to go beyond one in which campesinos are invited into the work of development institutions (public *and* private), and is to become a participation in which those institutions become more accountable to campesinos.

'Participation' is a notoriously malleable word, and in the hands of some, co-optation can be called participatory. In a review of participation in ATD, Biggs (1989) identified four degrees of participation which range from simply planting a trial on a farmer's field (contractual on-farm research) to situations in which the farmer and researcher treat each other as equals and farmers' ideas become one of the starting-points for ATD work (collegial on-farm research). Nevertheless, even in this collegial mode, the campesino has no means of holding the researcher, or the institution for which she or he works to account. For this to occur, campesinos need some sort of influential organization to represent their concerns.

Via this theme of accountability, the question of participation in ATD can be linked to the wider issue of rural democratization. In an editorial introduction to one of the few collections on the topic, Fox comments that rural democratization 'Involves the emergence and consolidation of social and political institutions capable of representing rural interests via-à-vis the State' (Fox 1990a: 1). For some commentators, such as Fox, non-membership NGOs are one such institution with the potential to represent and defend rural interests: '[t]he emphasis of [churches and NGOs] on changing political attitudes is crucial to any democratic project' (ibid.: 8–9). Lehmann (1990) makes a similar point: that the development, lobbying and research work of NGOs has played a critical role in strengthening the demands for citizenship rights that emanate from the grassroots. Friedmann draws attention to NGOs' ability to pressure the state to be 'more responsive to the claims of the disempowered' (Friedmann 1992: 158). For all these authors, the work of NGOs constitutes part of the process of building a more inclusive and equitable rural development.

Yet Lehmann's and Friedmann's invocations appear to draw more on their decades of experience in Latin American social change than on empirical support. Fox reminds us not to interpret such invocations as statements of fact: '[i]t should be noted, however, that the democratizing impact of religious and development organizations is more often assumed than demonstrated' (Fox 1990a: 9).

Furthermore, we must ask how far it is appropriate for non-membership NGOs to fulfil representative and lobbying roles. The self-proclaimed role of many NGOs is to strengthen the rural poor's capacity to make political demands for themselves, and to assume representative and administrative functions that the NGO may have performed at an earlier stage of a project. Their stated aim is to promote a process in which power is steadily transferred to the rural poor, with the NGO passing all decision-making and resources over to a membership organization. That at least is the theory. At such a point, it would become inappropriate for the NGO to presume to represent rural interests. Rural people could do that for themselves.

This point is taken further by those who emphasize that NGOs do not generally share the social and cultural origins of the rural poor they work with. These commentators argue that NGOs ought not be allowed the role of representing rural interests, for they have no justifiable claims to do so. Indeed, according to Rivera-Cusicanqui (1990) they may often misinterpret and misrepresent such interests. She criticizes NGOs working in highland Bolivia, where, she argues, they have operated with a liberal-western concept of democracy grounded in one-person one-vote that is inappropriate to what she termed the 'ayllu-democracy' in the Andean community, or 'ayllu'. Political administration and authority in these indigenous forms of social organization are, she says, based on quite different mechanisms for exercising accountability. Any claims that such NGOs might have made that they were promoting rural democracy by promoting elected local organizations of indigenous people would thus have been misrepresentations of the truth, says

Rivera-Cusicanqui. For these reasons, some argue that it should be membership organizations who lead local development administration, not other types of NGO (Annis and Hakim 1988; Blauert 1988; Kleemeyer 1991; CONAIE 1989).

These more critical observations on NGOs are important, but one senses that commentators, tired of criticizing the state, are now turning their guns on non-membership NGOs and overlooking problems in membership organizations. To do so may not only throw away the baby with the bathwater but also perpetuate a tradition of romantic populism which has proclaimed the virtues of rural peoples' knowledge, life and capacity to organise (Bebbington 1992; Richards 1990). Indeed, there is more than twenty years of literature suggesting that, if NGOs are not accountable to the grassroots, neither are the membership organizations of indigenous and peasant farmers. From Landsberger and Hewitt's 'ten sources of weakness and cleavage in Latin American peasant movements' (Landsberger and Hewitt 1970) to Carroll's finding that non-membership NGOs were more responsive and accountable to the beneficiaries than were membership support organizations (Carroll 1992), the message is clear: that the question of accountability is a vexed and complicated one, and that we cannot say *a priori* that any one type of organization is inherently more or less responsive to, or representative of, the needs of the rural poor. Indeed, those indigenous people who criticize the paternalism of NGOs, often also recognize that their own organizations are often controlled by largely unaccountable leaders (Bebbington *et al.* 1992a).

Aside from concerns about the limits on accountability within membership organizations, there is also much evidence they are administratively and professionally weaker than NGOs. In an evaluation of thirty different projects, Carroll (1992) scored NGOs higher than MSOs on the counts of effectiveness, wider impact, innovativeness and poverty reach; and Lehmann observed that:

> popular organizations tend to be small and financially dependent, and to have difficulty in sustaining a long life span, being plagued by fluctuating membership and unstable leadership. . . . Excessive faith in the managerial potential of grassroots organizations may breed disillusion both among intended beneficiaries of their activities and also among those who pay the bills. (Lehmann 1990: 200)

A review of ethnic organizations in Ecuador similarly demonstrated that they can suffer from administrative weakness, and that their development proposals tend to lack innovation (Bebbington *et al.* 1992a).

Where do these observations lead us? It ought surely to be the goal of a participatory rural development that membership organizations take more political and administrative control of ATD and other rural processes. Yet willing this to happen and be effective is not enough. After decades of exclusion, impoverishment and clientelist politics, it is not surprising that rural organizations suffer democratic and administrative shortcomings. Much capacity remains to be built.

While it may seem elitist and paternalistic to say so,[7] there are, then, good reasons to expect that NGOs will in the short and, perhaps, medium term have more impact than MSOs on poverty alleviation, participatory ATD and on policy change. This, however, is not an either/or choice: NGO or MSO. The point instead is that NGOs can have an important role to play in the political strengthening and administrative modernization of MSOs so that in due course they exert more effective direct pressure on government. They are able to provide and facilitate access to the sorts of technical and professional support and training that MSOs are often unable to provide themselves or gain easy access to. As the MSO matures, so one or another NGO can assist with more advanced forms of support: just as businesses require increasingly sophisticated support as they develop, so will MSOs and the rural poor in general. As 'outside' parties, NGOs may also be able to counterbalance the tendency of interest groups or individuals to 'capture' and dominate MSOs (Carroll 1992).

NGOs can, then, exert influence over NARS and ATD via their role in rural civil society. They can, we argue, act as pressure and advocacy groups on behalf of the poor, and they can build capacity.

NGOs, NARS AND AGRICULTURAL TECHNOLOGY: CONCEPTUAL TOOLS FOR EMPIRICAL ANALYSIS

These discussions of appropriate technology development, and of local capacity building lead us to similar conclusions: (1) that NGOs may have particularly useful contributions to make to a more equitable and participatory form of agricultural development; (2) that though some of their input involves more conflictive interaction with government, their contribution might be enhanced through a more organized and operational relationship with government institutions; (3) that little is known of quite how such relationships might work; and (4) that in all these discussions judgement of NGO contributions is based more on rhetoric than empirical study. This implies that much more needs to be documented regarding NGO actions in ATD and their past linkages to government. The case studies at the core of this book constitute part of that documentation.

However, to find a way through the case study material and address these issues, a conceptual framework is essential. We have selected a framework that aims to be appropriate for the questions being asked: namely, what are the actual and potential relationships between NGOs and government in ATD, which relationships might be usefully turned into operational linkages, how might these be managed, what are the contextual factors that impinge on these relationships and what might be some of the trade-offs involved? Our framework also reflects our concern to facilitate joint reading of this and the

ISNAR studies and to highlight certain areas in which we feel those studies were deficient.

Sub-systems and functions in the agricultural technology system

The central idea around which this book is organized is an adaptation of the concept of *agricultural technology system* (ATS), drawn from ISNAR's studies. As a heuristic construct, this concept can help us understand the way in which the public sector and the NGOs have performed different roles in ATD. This ATS can be defined as comprising all the individuals and groups working on the development, diffusion and use of new and existing technologies, and the relations between – and actions of – these individuals and groups (see Kaimowitz *et al.* 1990: 231). These organizations include, among others, NGOs, the NARS, commercial companies and farmer groups; individuals comprise members of these groups, such as individual extensionists, farmers, researchers and fertilizer sales people, and so on. The relationships between them are not only functional but also social. That is, they involve and often reproduce differences of power between actors and contribute to the definition of the identities of the individuals and institutions involved. When the nature of the relationship changes so might the identities and self-perceptions of the actors involved – an issue that has become very important to NGOs who were once in conflict with government and now find themselves collaborating with it.

In Figure 2.1, the ATS is represented in simplified form in terms of sub-systems which are composed of individuals and organizations engaged in particular activities with specific goals.[8] Each of these sub-systems carries out a specific function in the ATD process and each one potentially involves a group of actors. The *basic research sub-system* develops new knowledge, and the *applied research sub-system* develops new technologies on the basis of this knowledge to tackle certain concrete agricultural problems. The *adaptive research sub-system* effects changes in the technologies to adapt them more to agroecological conditions and to the socioeconomic realities of specific regions and producer groups. The *technology transfer sub-system* facilitates the adoption of technologies by the users. Finally, there is what we refer to as the *sub-system of technology use*, within which technological practices are implemented at the level of the campesino production unit. These practices may or may not incorporate the technologies developed by other actors in other sub-systems.

Although the idea of a system, and of its component sub-systems, is an artificial construct, we were concerned to present an analysis that, while being frank about the political economy of ATD, would also allow us to consider how functions were performed and actions carried out in generating and transferring technology. This was essential if we were to be able to make recommendations and prepare a piece of work that gave 'entry points' to managers, policy-

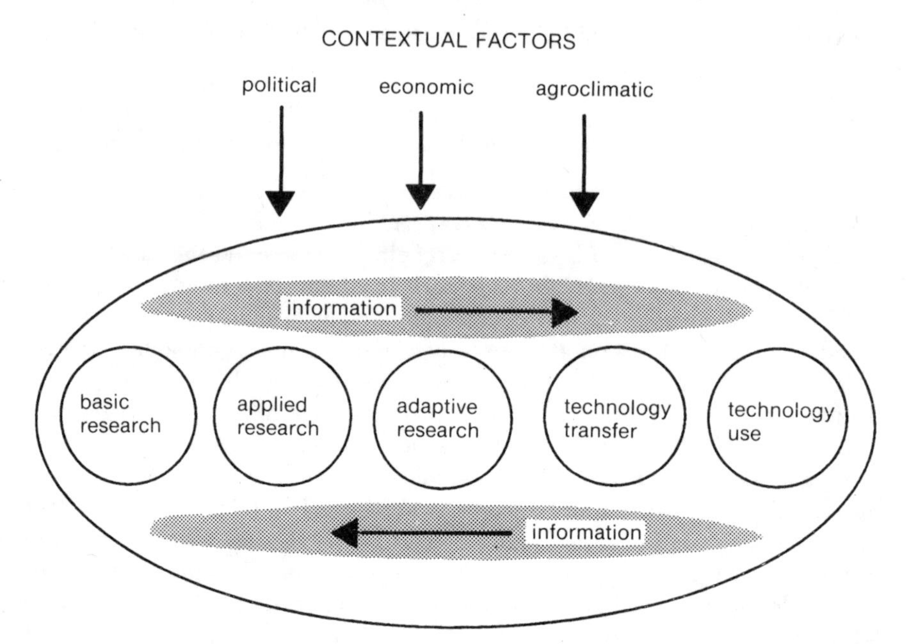

Figure 2.1 The ATS and its sub-systems

makers and other actors. We therefore chose this conceptual framework, the main elaboration being that we placed a heavier emphasis on what ISNAR called 'context' in order to stress the importance of politics and economics in shaping the system.

Interactions and linkages between actors in the ATS

The ISNAR studies tend to emphasize collaborative interactions between actors in the ATS, calling these linkage mechanisms (see below). However, there is no reason why interactions should be collaborative, nor why actors should wish to enter them. Indeed, as we have noted, NGOs may engage in more conflictive interactions with NARS, pursuing roles as advocacy and pressure groups. These more conflictive interactions are generally oriented to changing the goals towards which NARS operate in different sub-systems of the ATS or to changing the political and economic context of the ATS.

In other cases, NGOs and NARS might interact in order to communicate lessons and ideas, but without engaging in any formal operational collaboration.

However, for the sub-systems which make up the ATS to operate in a particularly effective and mutually reinforcing manner, organizational procedures should exist to facilitate the interaction of the different actors involved.

We refer to these organizational procedures as *linkage mechanisms* (cf. Kaimowitz *et al.* 1990). Linkage mechanisms should facilitate the two-way exchange of technological information throughout the ATS and enable the coordination of activities between actors in different sub-systems. Managing, manipulating[9] and negotiating them is thus a task requiring skill, sensitivity and a capacity to negotiate.

While there has been an important role for more conflictive NGO–NARS relations in the past, and there may still be to some extent in the present, our overall feeling is that in the future most advantage will be gained by more coordinated interactions; we have therefore paid particular attention to these linkage mechanisms in this study. Nevertheless, whilst linkage mechanisms are the key to improving the functioning of the ATS, they are also mechanisms through which relationships of power and accountability are exercised and can be changed. They therefore still offer an environment in which NGOs might be able to exert pressure on government, and their negotiation constitutes one means of furthering the democratization of ATD. For instance, a linkage mechanism in which campesinos and NGOs gain formal representation and decision-making power in part of the NARS would be a step towards such democratization, probably increasing, at the same time, the efficiency of research and extension activities by improving communication. One possible effect of such a linkage mechanism would be that in the applied research sub-system, technology more appropriate to campesino needs would be generated.

Individuals and institutions: actors and social relationships in the ATS

The different individuals and groups in the ATS can operate within one, some or all of these sub-systems. We refer to these individuals and groups as *actors*. It should be stressed that relationships within the ATS are always between actors, and not between sub-systems, although a relationship between actors has a direct impact on the functioning of, and the relations between, the different sub-systems. In this study we are mainly concerned with NARS, NGOs and farmers as actors, although development projects and other public entities, for example, do appear.[10] Theoretically, these actors can operate in every sub-system, although as we shall see, different actors function with different degrees of intensity in the different component parts of the ATS.

On the whole, we refer to organized actors, but organizations are made up of individuals who have a certain latitude to act with objectives and motives which differ from those of the institution itself (Giddens 1984; Long 1990). This is important, for it means we cannot conceptualize the relationship between institutions without also considering the relationships between individuals in those institutions. Individuals arrive in an institution with a history of particular social contacts grounded in their geographical, social, cultural and gender backgrounds, and with experience related to activities outside agriculture (e.g. political parties, educational institutions, etc.). Thus, for example, friendships between individuals formed in these contexts can facilitate relationships between institutions which are officially, at an institutional level, on bad terms. This can mean that a person in an NGO linked to the political opposition may frequently have family within government, even where, formally, the NGO has no contact with the state. To illustrate the point with a more concrete example from the research, one of the early factors in a collaborative relationship between one NARS and an NGO was that senior officials in each had children attending the same school (indicative of the similar social backgrounds of the individuals involved). While at that time the NGO and the NARS enjoyed no formal contact, a chance meeting between the two at a school event was one factor that led to discussions which ultimately led to a more formal collaboration. This example leads to a second important reason for recognizing the social contexts of individuals who make up NARS, NGOs and other institutions, namely that it alerts us to the class and ethnic structures within which such organizations are grounded. This is not to say that the institutions are simply agents of certain class interests, but it is to emphasize that they are scarcely neutral in class or ethnic terms. Thus, for instance, in Andean societies that are multi-ethnic, NGOs and NARS are largely composed of individuals of non-Indian backgrounds. While NGOs may claim to be more participatory and sensitive to Indian concerns, they cannot escape the fact that they are much more socially similar to NARS than they are to campesinos. This may facilitate NGO–NARS linkages, but may have negative consequences for NGOs' relations to campesinos, as we noted Rivera-Cusicanqui (1990) argued. Individuals bring to NGOs the social and cultural baggage of the classes from which they originate: these are in general classes that have long dominated and patronized the rural poor.

Contextual factors

The functioning of the ATS is subject to a series of contextual factors: political, economic and agroclimatic. Our heuristic separation of ATS from context means that, by definition, these factors are not strictly part of the ATS. None the less, they have a great effect on the intensity, the level and the quality of relationships between the various actors, and also on their actions and perspectives relating to ATD. These contextual factors have an impact both at the level of the whole system, and at the level of each sub-system. For example, political history and the extent to which individuals and NGOs experienced state repression in earlier periods will have much influence over the quality of NGO–NARS relationship in the more functionally grounded ATS; fundamental disagreements between NGOs and government over economic policy may also adversely affect NGO–NARS relations. Similarly, at the level of the sub-system of technology use, macroeconomic factors, price and credit policies, and agroecological conditions affect the possibilities for campesinos to incorporate technologies developed by other actors in the ATS.

Although we have put greater emphasis on the intervention of actors in the different sub-systems, actors also follow certain strategies to influence the contextual factors. Indeed, as we have already suggested, an important characteristic of NGOs is that they combine actions aimed at the ATS sub-systems with others which seek to influence contextual factors (e.g. promoting sociopolitical change and local democratization). In this sense, NGOs are agents of both social change and service delivery. They combine technical assistance with, for instance, work aimed at strengthening campesino organization, at lobbying for policy and political change, and so forth. It is characteristic of NGOs to argue that such technical and social interventions must necessarily be combined. In doing so they themselves insist on the relevance of contextual factors to the ATS.[11]

The ATS: a comment

Systems are frequently analysed out of their historical context. This leads to a focus on functional relationships between constituent parts of the system, which in turn can lead to recommendations that are inappropriate given the particular historical and social context of the system. We believe that many analyses of, and proposals for, change in NARS and agricultural technology systems have done just this, and as a consequence have led to neat systems diagrams of boxes and arrows with proposals that look perfect on paper but for social and political reasons are not feasible.

Whilst our use of terminology such as 'systems' and 'linkage mechanisms' may open this book and its recommendations to the same criticisms, our intention is to find a middle ground that tries to make policy and operational proposals that are meaningful to research managers, donors and NGO directors, while not being politically naive. At the same time, to the more academic and analytical of our readers, we hope to show both that sociopolitical and economic contexts have impinged on the form and nature of the ATS, but that at the same time institutions continue to act and interact in its different sub-systems. These interactions and activities offer policy and operational decision-makers 'room for manoeuvre', if not autonomy. It is this concern for the 'room for manoeuvre' of decision-makers that leads us to use a framework that is comparable with ISNAR's; by the same token, it is our concern to avoid some, if not all, of the simplifications of functional analyses that leads us to spend time emphasizing the social context of agricultural development and the institutions that foster it.[12]

CHAPTER SUMMARY

We have argued in this chapter that a reorientation of research systems to the needs of campesinos can be justified on several grounds that go beyond simple equity. Programmes of poverty alleviation, good government, sustainable development and not least economic growth would all benefit from such a reorientation. We then suggested that NGOs could help bring about such a reorientation through both direct collaborative interactions with government as well as through their wider actions in rural civil society – particularly through strengthening grassroots organizations ('building capacity') and through acting as pressure and advocacy groups. However, we suggested, there is still much to be learnt about how effectively NGOs do and could make such contributions.

We then laid out other concepts regarding NGO–state relationships that inform the argument of the book. NGOs can interact with government in both collaborative and conflictive modes. In much of their history in Latin America, this interaction has been conflictive both in the general political arena and in ATD in particular. This will be discussed in Chapter 3 in more detail. However, as we shall discuss in Chapters 3 and 4 a series of political and technological changes have created opportunities for more collaborative interactions. Taking full advantage of these (as well as knowing where best *not* to enter them) needs, however, more detailed consideration of the actual capacities, and comparative advantages of NARS and NGOs. These are discussed in Chapter 5. On the basis of this, more careful discussion of the nature of linkage mechanisms is necessary in order to inform institutions about if and how to engage in forms of coordination. These are the themes of the remaining substantial chapters of the book. To this substantive material we now turn.

NOTES

1 These two studies, one on on-farm client-oriented research in NARS, the other on research–extension linkages, were coordinated by ISNAR (International Service for National Agricultural Research).

2 'Although economic reactivation can occur without campesinos, and be based simply on stimulation of the modern agricultural sector, the preservation of democracy requires the incorporation of campesinos and other members of the informal sector into society' (de Janvry *et al.* 1989: 136). Recent protests in Ecuador, Venezuela and Peru testify to this.

3 None the less, this is not always so. Poverty and survival strategies may often lead the rural poor to practice non-sustainable resource management (Chambers 1987; Bebbington 1990, 1992; Woodgate 1991).

4 In Bolivia, campesino families with landholdings of less than 5 hectares cover 90 per cent of their own subsistence needs, and 50 per cent of the food requirements for population centres of over 10,000 inhabitants (Muñoz and Dandler 1986; Blanes 1985). In Colombia, campesinos produced 32 per cent of the total value of agricultural production and 43 per cent of the value of non-coffee production in

1988 (Machado 1986; Arango 1989). In Chile, although the relative contribution of campesino agriculture to the gross agricultural product has dropped, it still constitutes 25 per cent of that gross product, and this figure is higher in the case of some crops e.g. 31 per cent in wheat production (Gómez and Echenique 1988).

5 Another economic role of campesino production units should also be stressed. Campesino agriculture absorbs the majority of rural workforce and offers an important source of employment for rural people, especially given the current restrictions in urban labour markets.

6 The studies did, however, draw on experiences from Latin America.

7 'The strengthening of GSOs may be labelled elitist by some, but it is probably the most effective way of improving management and achieving greater stability.' (Lehmann 1990: 200).

8 Although we use the terminology of systems we wish to avoid functionalist explanations. Systems do not exist *in order to* carry out functions; rather, forces in society have created them in order to perform certain tasks. Similarly, systems do not exist as such, and do not have their own goals. Individuals and institutions (themselves composed of individuals) have goals and it is their pursuit of these goals that ultimately fashions the ATS and the relationships between sub-systems. Our emphasis on political economic context reflects our attempt to stress this social context. None the less, the terminology of systems is helpful to structure our thinking and the planning of interventions in ATD. This is why we chose to use this terminology.

9 We use the term 'manipulate' not in a pejorative sense but to convey the idea of skilled handling and negotiation of such relationships.

10 We recognize that the omission of the private-for-profit sector is a weakness of the study.

11 As do many NGOs, we place more emphasis on contextual factors in our analysis than do the ISNAR studies (see Ch. 3).

12 There is an emerging theoretical literature in the sociology, geography and anthropology of development that informs our approach, though we do not discuss it here. The writings of actor-oriented and structurationist approaches have been particularly helpful. See, for example, Booth, forthcoming; Long 1988, 1990; Long and van der Ploeg, forthcoming; Bebbington, forthcoming.

3

NGOs, THE STATE AND RURAL DEVELOPMENT

Changes in sociopolitical context since the 1960s

The main aim of this chapter is to analyse, in general terms, the sociopolitical context within which NGOs have emerged in agricultural development over the last thirty years in Latin America. This analysis will provide a basis for understanding their institutional characteristics, their perspective towards rural development, and the changing dimensions of their relationship with public institutions.

In Chapter 2 it was noted that the characteristics and activities of actors within the ATS, along with relations between them, depend on many contextual factors. This chapter will deal in detail with the influence such contextual factors have on the structure and functioning of the ATS. The intention is to point to broad tendencies in socioeconomic and political change that have taken similar form and had similar effects in the five countries of this study (Bolivia, Chile, Colombia, Ecuador and Peru). In particular, it focuses on the principal changes in agricultural and rural development policy (Table 3.1).

In schematic terms, this chapter argues that as the nature and size of the NGO sector has developed over time, successive stages in this evolution can be broadly identified. Different kinds of NGOs have emerged in each stage, at the same time as there have been changes in the nature of existing NGOs. Similarly, the state and its institutions have changed and these changes had important impacts on the NGO sector.

Many of today's strongest NGOs were founded in political contexts where electoral democracy was either suspended (sometimes with repression) or highly imperfect. In many respects, these NGOs kept alive the vision of a more democratic and participatory form of development at a time when there was little scope to pursue such goals through the state. Given the political context, they pursued this vision without much contact with public institutions. However, since the end of the 1970s there have been changes in this wider political context leading to the emergence of new types of NGO, a rethinking of roles on the part of existing NGOs, and more specifically to changes in conceptions of the relationship between NGOs and the state. Certain factors appear to favour closer relations. Paramount among these are the re-establishment of democracy (albeit still unstable) and the process of

Table 3.1 Dates of political and policy changes significant for NGO–state relations

	Periods of military government since 1960[a]	*Dates of major redistributive agrarian reform legislation*[b]	*End of redistributive agrarian reform*[c]	*Most recent transition to electoral democracy*	*Year of creation of Social Investment or Emergency Funds*[e]
Bolivia	1964–6, 1969–79, 1980–2 (1971–8, 1980–1)	1953	1970	1982	1986–9 Emergency Fund 1989– Investment Fund
Chile	1973–90 (1973–90)	1962, 1967	1973	1990	1991–
Colombia	None	1961	1972	1986–90[d]	
Ecuador	1963–6, 1972–8/9	1964, 1973	1979	1978–9	
Peru	1962–3, 1968–80	1964, 1969	1976	1980	1991

Notes

[a] Not all military governments have been violently repressive, though all are authoritarian and unelected. Particular repressive periods are suggested in parentheses.

[b] From Grindle 1986.

[c] These dates are our interpretations, several drawn from de Janvry (1981: 206); some coincide with clear policy reversals and new laws.

[d] Colombia had elected governments prior to this, but real steps to democratization came in 1986 with the end of the National Front governments, and in 1990 with legislative and constitutional changes.

[e] The date of the creation of Social Funds is something of a proxy for the date of public-sector reforms of particular significance for NGOs, although state withdrawal from activities had begun earlier.

Sources: Reid 1984; Dunkerley 1984; Ritchey-Vance 1991; de Janvry 1981; Grindle 1986; Wurgaft 1992.

public-sector reform, which has led the state and donors to look for new types of relationship with NGOs and civil society. Although many tensions continue to exist in state–NGO relations, this new context offers NGOs the possibility to augment and institutionalize the scope of their contacts with the state. Indeed, in some respects the changes seem to demand that they do so. However, the differences between types of NGO implies that they will not all respond in the same way to these challenges.

This chapter begins with an outline of policy and institutional changes in the public sector which have affected rural areas and the NARS. Following this, several factors which were instrumental in the development of the first generation of NGOs in the 1960s are discussed. We then move on to consider the impact of subsequent political and economic processes on the NGO sector and its relations with state institutions. The chapter concludes with a discussion of the processes of democratization and structural adjustment and a consideration of what they imply for easing prior tensions in state–NGO relations and creating more scope for more fruitful interactions.

It is important to recognize the risks inherent in such generalization, for these processes have not been identical in all the five countries. There are differences, for instance, in the strength of the NARS, in the forms land reform took, and in the universe of NGOs, which, moreover, in each country is itself varied and diverse. It is therefore difficult to generalize and undoubtedly some commentators will argue that we have ignored other key factors necessary for an understanding of one country or another. We acknowledge these drawbacks, and point to differences between countries when they are particularly important. (Statistical data in Annex 3 show similarities and differences between the five countries as regards selected macroeconomic and agricultural development indicators.)

STATE POLICIES AND AGRICULTURE BEFORE THE 1980s

Prior to the 1930s, agricultural research in Latin America was (when it existed) largely a private-sector affair, restricted to addressing particular problems in particular crops. Pardey *et al.* (1991: 244) suggest the slow development of state-sponsored ATD was an effect of the weak governmental structures that characterized the period, the fact that land and labour were relatively abundant – reducing pressures of demand for productivity enhancing technologies – and the lack of colonial heritage of agricultural research (unlike in Africa, for instance: Richards 1985).

By the 1930s, however, pressures for technological innovation were increasing. Concerns over competitive advantage were intensified with the Depression of the 1930s as markets in the industrialized world collapsed. At the same time population growth increased pressures on land and domestic demand for foodstuffs. In response to these pressures, governments began to establish units to promote or conduct ATD (Pardey *et al.* 1991: 245). This was the beginning of a gathering process of state intervention that ultimately led to the creation of institutions today recognized as NARS.

Alongside these government interventions, the involvement of certain international agencies had an equally important influence, and indeed the NARS tended to develop stronger contacts with key international donors than with other national institutions (Martínez Nogueira 1990). Referring to this period, Barsky speaks of the 'predominant influence of these funding donor bodies' (Barsky 1990: 114), one that has continued to the present day (Pardey *et al.* 1991a). The Rockefeller Foundation in particular was instrumental in the development of various national and international agricultural research services (Jennings 1988; Pardey *et al.* 1991). Its work in Mexico, oriented towards enhancing yields in food crops, led it into similar cooperative agreements in Chile, Colombia and Ecuador. It was also involved in the creation and development of the three international agricultural research centres based in Latin America.[1] As a consequence it has been identified closely with the Green Revolution and also, in the eyes of some, with the promotion of an approach to agricultural development in which technology and science (above all positivist science) are seen as the solutions to rural poverty (Jennings 1988). Jennings (1988) indeed takes this critique further and argues that the Foundation was instrumental in promoting a certain way of conducting agricultural science and rural development – a way in which social explanations of rural poverty were minimized and often excluded, to the advantage of explanations (and solutions) that focused on the technological backwardness of the rural poor. Either way, whether due directly or not to Rockefeller's work, a culture of seeking technological fixes to social problems has developed and persists in public agricultural research, and indeed has become one factor frustrating NARS' relations with NGOs – which endorse the importance of linking social and technological solutions.

The United States Agency for International Development (USAID) was similarly important in this period, establishing programmes in several countries (including Peru) which aimed to promote the institutionalization of a research, extension and education model based on the experience of the 'Land Grant University' in the United States (Jennings 1988; Pardey *et al.* 1991). In this model research centres (like those supported by the Rockefeller Foundation) adapted and then released technologies which already existed in the industrialized countries. According to Vessuri (1990), the creation of the NARS in Ecuador, Peru, Colombia and Chile were examples of the institutionalization of this model. Over time, these research centres and extension services grew, creating networks of public-sector agricultural development institutions across these countries (Box 3.1).

Together with the growth of the NARS, the state developed another series of institutions in this period for programmes oriented towards a more fundamental restructuring of the agricultural sector. This was a government response to a widely held analysis, put forward by institutions such as the Economic Commission for Latin America (CEPAL) and the Inter-American

Box 3.1 Public agricultural research institutions

Ecuador:	National Institute for Agricultural Research (INIAP)	1959
Peru:	Agricultural Research and Development Service (SIPA) – today known as the National Institute for Agricultural and Agro-industrial Research (INIAA)	1960
Colombia:	Colombian Agricultural Institute (ICA)	1963
Chile:	National Institute of Agricultural Research (INIA)	1964
Bolivia:	Bolivian Agricultural Technology Institute (IBTA)	1975
	Centre for Tropical Agricultural Research (CIAT)	1975

Committee on Agricultural Development (CIDA), which argued that the traditional agricultural sector had become the Achilles' heel of national development. CEPAL and CIDA recommended the abolition of the system of rural estates (or *latifundia*), the modernization of the campesino sectors and their integration into the national economy. Many argued this was an essential measure for both agricultural, and indeed national, modernization and growth. In pursuit of these changes, the state increasingly assumed a leading role in the development of rural areas and of the national economy (Carroll 1964).

The most far-reaching activities in this regard were the agrarian reform programmes of the 1960s and 1970s, which also were promoted by the United States through the Alliance for Progress. The aims of these programmes embraced economic modernization and political stabilization, and were particularly concerned to resolve the increasing social conflict in the Latin American countryside in the 1950s and 1960s (Stavenhagen 1970). To the extent that these reforms gave campesinos land, the programmes had the dual effect of increasing the number of farmers requiring services and making this farmer population far more diverse than it had previously been. Research and extension services had to begin to respond to their demands at the same time as maintaining their support to large and medium producers who had retained their land. Consequently, the NARS and other agricultural development institutions came under pressure from different types of resource-poor producers who needed and were capable of using different types of support and different technology, and who farmed in quite different and usually much more complex agroecological environments (Martínez Nogueira 1990). Similar pressures increased simultaneously in other sectors of the economy and their cumulative effect in the most extreme cases, such as that of Chile, was disastrous, leading to the violent response of 1973. In the less extreme cases, these divergent pressures caused problems in institutionalizing and consolidating programmes with a pro-campesino approach, given that such programmes were periodically subject to political attack (Box 3.2). The effect in all cases was that the public institutions and NARS tended to maintain a bias in favour of large and medium producers (CAAP 1991), while at the same time they became more complex and bureaucratic (Pardey *et al.* 1991).

Box 3.2 Conflict and complexity in NARS

'The state became more like a confederation of public institutions, each one with different analytical and operational capabilities, and all subject to pressures from corporate interests' (Martínez Norgueira 1990: 96).

The redistributive policies of agrarian reform were, in any case, paralysed during the 1970s: in Peru, between 1975 and 1980, in Chile in 1973, in Colombia in 1971–2 and in Ecuador formally with the 1979 Agricultural Development Law. In some countries (Bolivia and Chile) the freezing of these progressive reforms occurred under authoritarian and repressive regimes which subsequently ceased to channel significant resources towards rural development programmes. In other cases, such as Ecuador and Colombia, agrarian reform programmes were unravelled without such intense repression, to be followed by policies of integrated rural development (Barsky 1990; Grindle 1986). These policies set out to coordinate technology transfer, credit provision, campesino organization, the provision of infrastructure, health care and myriad other activities. One effect of integrated rural development policies was to increase yet further the complexity of the institutional structure and organization of public-sector agricultural development activities.

Overall, agrarian reform and rural development programmes tended to have centralizing effects, and as a result were often poorly adapted to local conditions (Barsky 1990: 114). They concentrated decision-making power, resource, and agricultural development activities in the public sector and fostered bureaucratic complexity. This led to many inefficiencies, which have attracted criticism from NGOs and donors alike, and have subsequently been the object of programmes of public-sector reform and shrinkage (Box 3.3) (CEPAL 1990: 35). The centralization of resources within the public sector also tended to attract attempts by different interest groups to influence public planning and so influence the distribution of these resources (cf Kruger 1976). For example, groups of large-scale farmers tried to influence the research perspectives of the NARS and politicians tried to gain control of development programmes in the hope of generating support and votes (Grindle 1986).

In addition to these inefficient and largely undemocratic procedures in rural development, this period was characterized by authoritarian regimes. Electoral democracy was weak and short-lived at best, and often totally absent – as were political liberties. Only in Colombia was there civilian government throughout

Box 3.3 Some tendencies of state policies criticized by NGOs

- Growth, inefficiency and bureaucratization of the public-sector apparatus in rural development.
- Institutional centralization of power and weak campesino participation.
- Bias of NARS and other institutions towards wealthier farmers; NARS ignore campesinos.
- Weakness or absence of democratic procedures in the definition of national policies.

this period and even here democracy was more formal than real (Ritchey-Vance 1991). In Bolivia and Chile the suspension of democracy was violent and repressive. There were also periods of repression of campesino organizations, progressive intellectuals and the universities (Lehmann 1990; Velasco and Barrios 1989).

These factors – on the one hand state bureaucratization and inefficiency, and on the other, the weakness of democratic procedure in national and rural development – were of critical importance in the development of many NGOs founded from the 1960s onwards. They affected both the perspectives NGOs adopted, and the nature of their relationships with public institutions (Box 3.4).

Box 3.4 Factors in NGO development in the 1960s and 1970s

- The intellectual climate (liberation theology, Freire's ideas on popular education, community development and cooperativism, agrarian reform).
- Agrarian reform programmes:
  - generate expectations when launched;
  - create disillusion among some public officials when paralysed.
- Political repression and dictatorship.
- The support of a significant sector of the donors.

SEARCHING FOR DEMOCRATIC DEVELOPMENT: THE EMERGENCE OF NGOs

During the 1950s and 1960s, as public-sector agricultural development institutions began to consolidate, other currents of thought and action regarding development strategies were also emerging from within Latin American civil society. These currents had an important influence in the early development of the NGOs, and indeed NGOs subsequently carried them forward. Amongst the most important of these influences were (1) the growing commitment of the Catholic Church to the poor; (2) the ideas of Brazilian educationalist Paulo Freire, who developed means of uniting popular education and political empowerment (Freire 1970); (3) experiences of community and cooperative development; (4) contacts between development professionals, intellectuals and the new campesino movements of the 1950s and 1960s; and (5) experiences of and debates about agrarian reform and land tenure. On the basis of these different intellectual and political currents emerged conceptions of a rural development that would be more democratic and egalitarian than the programmes that had so far passed as development (see Lehmann 1990).

The 1960s marked a turning-point in the Catholic Church. The Second Vatican Council in 1965 and then the Congress of Medellín in 1968, provided the basis for what has become known as Liberation theology. Through these discussions, parts of the church developed the idea of an 'option for the poor' in which the church would work not only for the spiritual well-being but also the welfare and the political emancipation of the urban and rural poor.

Liberation theology, with its base Christian communities, which gave the laity access to the Bible and later became the basis of local self-help actions, combined the formation and consolidation of grassroots organizations with ideas of local democracy and Freire's ideas about the role of informal education in the formation of a critical consciousness amongst campesinos as part of the process of social liberation (Lehmann 1990). Some NGOs grew directly out of these tendencies within the church; some were more radical than others, but all were oriented towards social development work with campesinos. Among today's strongest NGOs, examples of these are CESA and FEPP in Ecuador, CIPCA in Bolivia, SEPAS in Colombia, and INPROA and IER in Chile (Box 3.5; Carroll *et al.* 1991; Gómez and Echenique 1988; Lehmann 1990).

Another important influence in this period was the community development movement and subsequently the agrarian reforms. Community development began with the assumption that campesinos had significant potential to enhance their production and welfare and that strengthening local organizations offered an important means of stimulating their capacities. Community development programmes therefore focused on promoting local organizations (with much interest in cooperatives), training leaders, education and on a participatory development process (Grupo Esquel 1989; Barsky 1984). Attempts were made to link these socio-organizational activities to technology transfer processes for the modernization of campesino agriculture. The results of this approach, which was adopted by institutions such as the Andean Mission, were however, disappointing. It failed to make a significant impact, and it became clear that more fundamental changes were required to address rural poverty and modernize campesino agriculture. The programmes ended abruptly towards the mid-1960s when US funding was cut. However, the programmes left behind them a body of methodology, theory and practical experience in grassroots participation in development which later reappeared as a prominent part of NGO philosophy.

During the 1950s and 1960s, campesino movements gained in strength, nurtured by growing demands for land and by the effects of community development programmes in leadership formation (Carroll 1964; Grupo Esquel 1989). Although the experience of these movements was variable, suffering as they did from many weaknesses and internal problems, they had great influence over the thinking of many professionals and intellectuals who worked with them, and who would later carry with them a commitment to the need for strong campesino organizations.

One of the effects of growing campesino demands for land was the decision to implement agrarian reform laws in an attempt to quell rural unrest. Despite the fact that the laws of the early 1960s were hardly implemented and, according to de Janvry (1981), represented the minimum effort by the state to tap into the benefits of the Alliance for Progress, they created expectations of a more equitable development. The first laws failed to dissipate social tensions in the countryside and as a result in some countries a second round of laws was passed; in others the application of the initial laws was intensified by radical regimes (e.g. Chile 1971–3; Peru 1969–74).

These various currents of policy thinking emerging during the 1960s shared the belief that development had to be participatory, ought to be based on campesino organization, and should link any technological intervention with corresponding broader social changes. These ideas influenced the thought of a whole generation of professionals and activists, some of whom created NGOs to pursue these aims. Others went to work in universities and political parties. But many others went to work for the state, and during the height of the agrarian reform and rural development movements, these public institutions attracted many young professionals committed to campesino development. In these periods, public institutions seemed to offer a vehicle to pursue their vision of a more egalitarian rural development, and the opportunity for interesting and innovative work (Cleaves and Scurrah 1980; Grindle 1986; Ritchey-Vance 1991).

However, with the end of land reforms, and the political upheavals and authoritarian regimes of the 1970s, the hope that such institutions could offer a viable mechanism for promoting a more egalitarian and democratic style of development all but died in the eyes of many. This was most resoundingly so in Pinochet's Chile, where agrarian reform and campesino development activities not only ended, but left-leaning professionals were sacked, and some disappeared (Gómez and Echenique 1988; Sotomayor 1991). Similarly, social sciences in the universities suffered financial cutbacks and repression, and leftist political parties were often outlawed. In the different countries and for different reasons, it thus became ever more difficult to conceive of the possibility of working for participatory development within orthodox institutions (state, universities, political parties). As a result, some professionals saw NGOs as an alternative vehicle for pursuing their goals (Boxes 3.5, 3.6 and 3.7). In cases such as Chile, they were also the only means of obtaining employment (Box 3.8).[2]

Box 3.5 The social origins of NGOs in the 1960s

The Ecuadorian Centre for Agricultural Services (CESA) was founded in 1967, influenced by ideas from the radical Catholic Church and community development programmes, at a time when campesino movements were demanding land. It was formed by leaders of several class-based organizations (amongst them today's National Federation of Campesino Organizations). CESA's initial objective was a pilot project of agrarian reform in the rural holdings of the Ecuadorian church. Its philosophy stressed that Ecuadorian society was characterized by the marginalization of campesinos from decision-making in national political and economic life. Consequently, it saw its main activity with the marginalized rural population as organization, so that they could be integrated in a more egalitarian manner into society. To achieve this organization, the NGO implemented agricultural training and service programmes.

The origins of the Pastoral Social Secretariat (SEPAS) in Colombia were more closely linked to changes within the Catholic Church. With the Council of Vatican II and the document 'The Church and Change' presented to the Latin American Bishops' Conference in Medellín, an increasingly explicit social commitment to the poor became evident in Colombia's church. The appearance of socially committed groups of priests, such as 'Golconda' and 'Priests for Latin America', also established the basis for the emergence of ecclesiastical NGOs such as SEPAS.

The origins of today's grassroots support organizations (GSOs) in Peru can be found in the religious, social and political movements within Peru over the last thirty years. The earliest organizations appeared in the 1960s, created and influenced by a circle of professionals linked to the Catholic Church's social action movement. They were influenced by the ideas about liberation theology of Gustavo Gutierrez, and those of the Brazilian educator, Paulo Freire, concerning popular education and *conscientización*. These early organizations were also influenced by Vatican II, which brought the Catholic Church much closer to the poor and led to a flurry of grassroots programmes.

Source: Carroll *et al.* 1991; CESA 1991; Gonzalez 1991.

Box 3.6 FUNDAEC: a university-based NGO

At the beginning of the 1970s, Colombian universities were passing through a critical period, as was the campesino population. In academic circles, new social movements began to question existing structures. In the countryside, however, campesino organizations were in decline and the agrarian reform was having little or no impact. In this context, a group of teachers at the University of the Valley created FUNDAEC in order to elaborate development proposals with campesino groups.

Source: FUNDAEC 1991.

Certain donors shared this disillusion with public institutions: in the most extreme cases, these agencies pulled out of certain countries in protest at the

Box 3.7 NGOs in Peru during the 1970s

In the 1970s, as the crisis of higher education deepened with increasing enrolments, declining resources, and increasing politicisation, a growing number of university professors sought to satisfy their needs for carrying out research and being involved in relevant social action by working in GSOs, which were relatively free from the pressures and chaos of the campus.

(Carroll *et al.* 1991: 97)

With the end of the Velasco reform period, the new administration began to dismantle and reorganise the government agencies responsible for implementing the reforms. From the many professionals who were obliged to leave came the founders of a number of important GSOs.

(ibid.: 98)

The majority of NGOs in Peru began their activities in the 1970s, very few actually began before 1970 but they all grew throughout the decade. In general terms, they were encouraged by the changes implemented by the military government of the period, and in particular by the agrarian reform and by the possibilities for social action that this opened up.

(Torres 1991: 4)

Box 3.8 NGOs: survival strategies and policies under dictatorship

The [Chilean] dictatorship gave birth to institutions that resolve – and here we must be honest – not only the needs that social groups have, but also the problems of professionals – [those of us] who had no place to work; not only no place to work in the field we wished, but no type of work at all.

(Daniel Rey, AGRARIA; quoted in Loveman 1991: 10)

It is in the decade of the military dictatorships that both church-based and the few independent NGOs naturally came together as a means of protecting themselves against indiscriminate repression.

(Velasco and Barrios 1989: 1)

repression (as in the case of COTESU, the Swiss Technical Cooperation agency in Bolivia: Zeller 1991). Some agencies of course, already had a policy of not working with public institutions. Other bilateral donors and foundations which had worked with governments began to intensify their search for other institutional means to implement programmes with a social focus: NGOs appeared as one such potential vehicle. Of course, political and conceptual differences between the agencies continued. Some were made up of lay-people and others were more explicitly linked to the church; and they subscribed to different visions of democracy (Christian, social, socialist, etc.). Although the donor agencies therefore tended to support those NGOs which shared their outlook, in overall terms they played a crucial role in the growth of the NGO sector in this period.

Many of the larger NGOs currently operating in the five countries were set up during the 1960s and 1970s, and despite their different national contexts, they shared (and still share) a sense of common identity. Uniting them was the belief, formed during this period of intense political and intellectual debate and conflict, that they represented an attempt to protect and promote, at a local level, a different vision of a more democratic and participatory campesino development. The institution of the NGO offered a means of questioning the ruling political system and its development models. To a certain extent, this has been their *raison d'être*, and their source of institutional identity. As we shall suggest, whilst this may have inspired their vigour and determination to survive during these difficult times, it is also a factor that now complicates any attempts to begin working with the institutions of the political system that they were born in order to question.

NGOs AND THEIR RELATIONSHIP WITH THE STATE IN THE EARLY YEARS

In broad terms, the contexts within which NGOs arose can be divided into those characterized by regimes that were repressively authoritarian or moderately authoritarian. Although this distinction is hardly ideal, it draws attention to the difference between Chilean and Bolivian contexts, where military governments were harsh, to say the least, and other contexts where military regimes or (as in Colombia) elected governments were not nearly so harsh, although they remained far from innocent or democratic. The context influenced the aims of the NGO *vis-à-vis* the state: under moderately authoritarian regimes, many worked to change the *modus operandi* of public institutions in order to make them more participatory and oriented towards campesino reality (the case of Colombia: e.g. González 1991). In the more authoritarian cases, many NGOs worked in resistance to a repressive state and fought for the return to democracy (the case of Chile: e.g. Sotomayor 1991).

In the most repressively authoritarian cases, such as Chile from 1973 and Bolivia in the 1970s and early 1980s, the political situation was highly polarized and the NGO–state relationship was generally a hostile one. In a context where people were unable to find work because of their political backgrounds, NGOs offered the possibility of an income and provided a means of maintaining contact with the grassroots in a situation where leftist political and campesino organizations were repressed and where the state did little to respond to the material needs of the poor. At the same time as meeting an urgent social need, many of these NGOs were formed as a means of political resistance against the repression of civil rights, the shelving of social reforms and the suspension of democracy (Box 3.9: Landim 1987; Lehmann 1990; Velasco and Barrios 1989). Some viewed their work as an attempt to develop policies and methodologies which could be implemented on a wider scale by the state once formal democracy was re-established.

Box 3.9 A political commitment

'The [NGOs] have their raison d'être in their complete identification with the Bolivian popular movement, as a legitimate object whose aspirations, rights and needs should be met' (Velasco and Barrios 1989: 10).

Under these regimes, the work of many NGOs was difficult and some suffered repression at the hands of the state. At the same time, the most repressive governments often prohibited contact between public-sector agricultural institutions and NGOs. In Chile, for example, neither INIA nor INDAP (the Agricultural Development Institute) had any operational contact with NGOs (Sotomayor 1991).

None the less, even in these repressive contexts, there was a certain flexibility at the level of personal contacts. In this regard, it is important to stress a point made in the first chapter: institutional actors (NGOs, NARS, etc.) are made up of sub-groups and individuals. These individuals have a certain room for manoeuvre to act outside the limits defined by their institution's mandate, and in fact through such actions may contribute to changes in those very institutions. Even the dictatorships were unable to exercise total control over the work of the staff of public-sector institutions. Individuals in the public sector who sympathized with some of the aims of the NGOs or who wanted to pursue a line of work involving closer contact with campesinos therefore often found ways to satisfy these goals.

In some cases, then, there were contacts between NGOs and public-sector institutions, although these were generally informal. None the less, their occurrence complicates the stereotypical view that there was an unbreachable distance between the two sectors in such circumstances. Under the Pinochet regime, even when at an institutional level the government interfered politically with AGRARIA and GIA (viewing them as part of the political opposition) and kept them under surveillance, both NGOs in fact received informal support from individuals within the public sector (Aguirre and Namdar-Irani 1991; Sotomayor, 1991).[3]

In the less repressive contexts (e.g. Ecuador, Peru, Colombia), the state–NGO relationship was not as hostile, yet neither was it easy. Although NGOs perceived their work as a critique of public programmes, few saw it as a means of political resistance to the same extent as in Chile. Where there was a degree of access to the public sector, some NGOs researched and promoted alternative approaches to rural development. They criticized public-sector approaches to ATD on the grounds that they promoted inappropriate changes, were socially and technologically inefficient and were non-participatory. They hoped to demonstrate that other approaches to development were both viable and desirable. A degree of political freedom existed within which NGOs could promote these different models. Not all NGOs took advantage of this opportunity, but the political situation made it easier than it was in the more repressive situations. Where contact was sought with the government, this once again tended to occur at a personal, rather than institutional, level – although there were some cases of this. During the Velasco government in Peru, NGOs who viewed the government as a progressive regime collaborated with it. In Ecuador, some contact between NGOs (such as CESA) and agrarian reform institutions similarly occurred. CESA, for instance, cooperated in the sub-division of large properties and later on there was contact between it and the NARS (Box 3.10).

Box 3.10 Early relations between an NGO and a NARS

The institutional relationship between CESA and INIAP was established at the start of 1976, when Ecuador was still under military rule. The relationship arose for a variety of reasons: a shared interest in certain agricultural technologies; CESA's need to gain access to specialized technological and scientific resources; and the desire of CESA to ensure that INIAP's technology should be more available to campesinos. Relations in a formal sense between the two institutions were in general good, although their content and quality fluctuated.

Source: CESA 1991.

However, even in these more favourable contexts, it should be recognized that such NGO–public sector relations were rare. The political manipulation of public institutions, their unstable personnel and resource levels, and the bureaucratic hurdles that had to be negotiated in any inter-institutional agreements remained serious obstacles to collaboration.

In addition to these general political obstacles to NGO–state linkages, there were more cultural and even epistemological ones in the specific case of ATD. These stemmed from the dominance of positivist and technocratic approaches to agricultural development in the public sector – a dominance that, as noted earlier, had been fostered by particular donor agencies. As Octavio Sotomayor of GIA explains:

> The NARS emerged in the 1950s, much influenced by North-American scientific and financial organisations. Inspired by the logic of the Green Revolution the outlook of these organisms is technocratic and productivist . . . this way of doing science creates difficulties for any relationship with NGOs which, given their origin and evolution, feel more at ease with social sciences and have a different vision of technical change. (Sotomayor, personal commmunication, 1992).

In short, the much talked of conflict between social and biological sciences in international agricultural research (de Walt 1988; Rhoades 1984) was also played out in Latin America as a tension between NGOs and NARS. It is, perhaps, a particularly significant source of tension because it will not easily be solved by political changes.

Thus the main legacy of the 1970s is the memory of difficult and distant state–NGO relations (Box 3.11). This legacy has constituted an important part of the NGOs' discourse. However, albeit not very common, we must recognize that even during this period, some NGOs did maintain informal but operational relations with the public sector. Even more significant is that, when they were founded, NGOs maintained the view that their work was directed towards the achievement of a new and democratic state. To some extent NGO projects represented attempts to draw up proposals for wider implementation in more amenable political contexts. The logical implication of their own reason for existing was that one day the NGO would work with and through government.

Box 3.11 Broad tendencies in the state–NGO relationship under different regimes

Under repressive authoritarian regimes, NGOs had infrequent contacts with governmental institutions. In many cases, they conceived their strategy as an act of political resistance, oriented towards ending authoritarian rule.

Under moderately authoritarian regimes, contact between NGOs and public-sector institutions was a little more feasible. NGO strategies were oriented towards promoting more participatory practices in the public-sector institutions, in order to make them more accessible to campesinos.

In general, if they existed, state–NGO relations prior to the transitions to democracy were weak. They were largely of an informal nature and were between individuals. None the less, it should be noted that even so, NGO strategy was very much based on the idea of a state: at the end of the day, their aim was to work towards a new, democratized and participatory state.

THE CONTEXT CHANGES: DEMOCRATIZATION, PUBLIC-SECTOR REFORMS AND THE PRIVATIZATION OF PUBLIC SERVICES IN THE 1980s

The question today is whether these more amenable contexts have arrived, or at least are close at hand. A series of changes in the sociopolitical and economic contexts of the five countries during the 1980s have confronted NGOs with the challenge of rethinking their relationship with public-sector institutions and their own role in society. At the same time, the same changes force public-sector institutions to rethink their role in development and their relationship with NGOs: as a result they are beginning to see NGOs as an institutional resource for implementing public programmes. According to some commentators, these changes imply greater possibilities (and a greater necessity) for a more constructive and coordinated NGO–public sector relationship (Breslin 1991). In this section we will concentrate on four main elements of these forces for change in the NGO–NARS relationships:

- Political changes, above all the return to electoral democracy and the tendency towards political decentralization.
- Public-sector reforms stemming from structural adjustment measures.
- The policies of the funding agencies.
- The proliferation of NGOs and the emergence of a new breed of more opportunistic, less radical NGO.

We want to suggest that, although these changes do not necessarily imply that NGOs must establish a relationship with public-sector institutions, they at least mean that they have to think far more carefully and impartially about the possibilities of such relationships.

Political changes and the NGO–public-sector relationship

Democratization

The 1980s saw the re-establishment of civilian governments and processes of electoral democracy: in 1979 in Ecuador, 1980 in Peru, 1982 in Bolivia and 1990 in Chile. Although the situation was different in Colombia, which had a civilian government throughout the entire period, it was only in 1986 that the National Front government ended, and in 1990 that 'after a century of virtual monopoly by the two traditional parties, the electoral landscape filled with players' (Ritchey-Vance 1991: 14).

With these changes, some observers have suggested that one of the main obstacles to an NGO–state relationship has been overcome, and that now is therefore the time for NGOs to start working more closely with public-sector institutions (Barsky 1990; Jordan *et al.* 1989). There are different reasons for such recommendations. The fact that these governments are elected supposedly means that they will be less repressive and more trustworthy. Some also argue that these are the governments NGOs have been waiting for and that NGOs should be true to the logic of their previous actions. If their previous work involved the elaboration of methodological and technical innovations for implementation on a wider scale by new, democratic public-sector institutions, now is the time for NGOs to seek the means for such a transfer of their proposals.

Furthermore, the fact that these countries now enjoy governments that are elected with a popular mandate, puts some pressure on NGOs which are unelected (or self-elected) bodies to rethink their relations with government. Given that NGOs receive resources to be used for the social good, some similarly insist that there should be mechanisms through which society can guarantee that such resources are used for these ends (Kohl 1991). Now that electoral democracy is in place, some (both within and outside the state) assert that government supervision could act as such a mechanism, and that therefore some relationship between the state and NGOs ought to exist. On grounds such as these, the Bolivian government proposed in May 1990 that all NGOs should be registered with the government and that state agencies should participate in NGO evaluations – both to maintain their accountability and to learn from their innovations. NGOs have been highly critical of this suggestion.

Undoubtedly democratization has created a more favourable climate for NGOs and has facilitated collaboration between NGOs and the public sector (de Janvry *et al.* 1989; Loveman 1991). This is recognized by many NGOs in this study. However, at the same time these NGOs emphasize that one cannot presuppose that electoral democracy necessarily leads to harmonious state–NGO relations or to more transparent government practices.

Other commentators have also noted that 'the return to democracy has often been more formal than real' (de Janvry *et al.* 1989: 135). It is feared that the democracies remain, as yet, unconsolidated and that to hand over much of the NGOs' autonomy to the state would be highly risky if authoritarian regimes came to power again. The political crisis in Peru, with the imposition in 1992 of an unelected form of government, suggests that this concern is justifiable. This has meant that mistrust of the state continues, particularly in those countries or places where repression has been more serious or where democracy is still very fragile.

It is also the case that the mere fact of being elected does not mean that a government will cease to repress or mistrust NGOs. Ideological differences and institutional competition both continue. Several elected governments have tried to control NGOs to a greater extent than would be necessary if all they were interested in was to learn from their innovations or to ensure responsible use of resources. The proposal for an NGO register in Bolivia, for instance, included another for levying a tax on NGOs, and was interpreted by some NGOs as an attempt on the part of the public administration (and the parties in power) to exercise control over the work of NGOs (many of whom have political sympathies that differ from the government's). An attempt to exert political control over NGOs was similarly apparent in Ecuador when the Leon Febres Cordero government (1984–8) harassed some NGOs. In Peru, the Belaúnde government (1980–5) was suspicious of NGOs, to the point of denouncing some of them as subversives. Under the subsequent APRA government in Peru, supposedly more sympathetic to NGO ideology, political competition also impeded relations between NGOs and public institutions (Carroll *et al.* 1991).

Thus, while the transitions to democracy have without doubt created more opportunity for constructive relations between NGOs and the state, these opportunities remain constricted.

Administrative decentralization and the NGOs

Another element in the reform of the political process and the public sector has been that of administrative decentralization. This has had objectives that are both political (to strengthen local democracy) and economic (to promote resource management that is more efficient and better suited to local conditions).

Under the new constitution and the municipal reform laws of 1986, Colombia is experiencing an important process of decentralization (Ritchey-Vance 1991: 36–9). This process has also had more specific implications for the administration of the ATS at a local level (see Chapter 6). In Peru there has also been a process of regionalization under which (gravely underfinanced) regional governments have become responsible for agricultural research and extension institutions (Box 3.12). Although Torres underlines the fact that, to date, state–NGO relations at a regional level have been very similar to relations at central level, he does note that regionalization has given rise to a situation in which 'regional societies are increasing their expectations of, and demands on NGOs' (Torres 1991: 5). In Chile, during the period of military rule, a significant degree of administrative decentralization occurred, although under the dictatorship this was not linked to a process of democratization. In the 1990s, however, this administrative decentralization is being steadily fortified with efforts to enhance local democracy, and locally elected governments now have powers to administer socioeconomic development programmes for their respective regions.

Such reforms have various implications for NGOs, and as examples in this study demonstrate, NGOs have shown an interest in the possibilities that could be opening up (Ritchey-Vance 1991: 38). In Chile, the Group for Agrarian Research (GIA) is putting special emphasis on analysing the potential for, and nature of, linkage mechanisms with municipal government, as part of its more general aim to influence and participate in the definition of local development policies (Sotomayor 1991). In Peru, both the Centre for Research, Education and Development (CIED) and IDEAS saw regionalization as opening up new possibilities for engaging in local development – especially given that a number of seats within regional government are to be selected from the institutions of civil society. Although these spaces for local administration in municipal and regional government could easily be dominated by local elites, they none the less offer to NGOs and campesinos the possibility of direct participation in the decision-making process in order to resist such domination. The experience of the NGO SEPAS (The Diocesan Secretariat for

Box 3.12 Public-sector cutbacks and regional decentralization in Peru: impacts on NGO–state relations

In Peru, various processes of change in recent years have had considerable impact on NGOs and their relationships with the state.

- The violent reduction in the state's capacity, with cutbacks in programmes, projects and public-sector wages has led to the departure of many staff. With this extreme weakness of government, large sectors in civil society have looked to NGOs to fill the gaps left by the state's withdrawal. This and other neo-liberal policies have led to something of an identity crisis in many NGOs, because the current situation is totally different from the context in which they initially emerged.

  This process has also led to a change in NGO–state relations. Until 1985 these relations were largely distant and characterized by mutual distrust. With the decline of the state, NGOs and the public sector have begun to work closer together, assuming joint responsibility in meeting emerging challenges. This coming together, although slow, has led to several positive results, so much so that the state is now channelling public resources via NGOs.
- In 1990, in this context of public-sector crisis, the state began a formal process of regionalization in which new regional governments have been created, and administrative responsibilities decentralized. This has engendered two particularly interesting initiatives:
  - At the end of 1990, various NGOs created an institution called SUMMA as an NGO-coordinating mechanism with the specific mandate of supporting the process of public-sector decentralization.
  - The Centre for Regional Andean Studies 'Bartolomé de las Casas' in Cusco and the Centre for Campesino Research and Development (CIPCA) in Piura have entered preliminary agreements with their regional governments and the National Institute for Statistics and Information to develop regionalized databases as a first step towards creating regional information systems.
- The combined effect of these changes has been that donors too are seeking increased NGO participation in projects, because the legitimacy of the state is now so weak.

Source: Torres 1991.

Pastoral and Social Work) suggests that much can be achieved by this sort of involvement in local politics (Box 3.13).

Aside from opening up new possibilities of political participation, these changes may also give NGOs greater roles in the implementation of local programmes. Of course, administrative decentralization need not imply that local governments will be given the resources, or revenue raising powers necessary to generate resources required to fulfil the new functions being handed over to them by central government. But, in so far as they can generate resources (e.g. on the basis of their ability to collect local taxes) to contract private agencies for implementing local programmes, NGOs will be able to take up such contracts (Jordan *et al.* 1989: 260; Reilley 1992).

Box 3.13 SEPAS and the municipalities in Colombia

One of SEPAS's objectives is to stimulate a regional development process which has as one of its central elements the development of a new and stronger form of municipal administration, that is more responsive to popular needs and more committed to a form of social development project which would build on and integrate the contributions of public bodies, NGOs and popular organizations. With this aim in mind, SEPAS has tried to strengthen grassroots organizations and has promoted relations between the municipalities and the provincial administration in order to give popular organizations more influence over public-sector institutions. SEPAS participated in the plans for the Development of Municipal and Provincial Government, and in the elaboration of agricultural base-line studies coordinated by the municipal government. It also maintains operational agreements with municipal agencies with whom it shares responsibility for the provision of rural education, agricultural-technology transfer, health care and irrigation infrastructure services. In the case of irrigation, SEPAS managed to establish a cooperative around water management. Membership of the cooperative cut across institutional boundaries bringing together forty grassroots organizations, SEPAS, four municipal governments and four parishes.

SEPAS recognizes that its attempts to promote change in municipal government have been strengthened by the administrative decentralization process in Colombia. The municipalities lack the necessary experience to provide the services that decentralization now obliges them to provide, and have therefore approached SEPAS to learn from and take advantage of its experience.

Source: Gonzalez 1991.

Even if decentralization processes are imperfect, they will, then, open up interesting possibilities for local NGO–state linkage. This is particularly significant, because our own research endorses the assertion of Bolivian NGO-watcher Godofredo Sandoval, who comments that the 'most innovative experiences of the [state–NGO] relationship do not occur between institutional directors, where prejudiced views still persist, but rather at local and regional level on the basis of concrete situations' (Sandoval 1991: 4).

Structural adjustment, the reduction of state capacity and the NGOs

The crisis of Latin American economies[4] led to a series of policy changes during the 1980s (the 1970s in Chile) which have been termed programmes of 'structural adjustment' – programmes revolving around economic liberalization, reduction of public spending and privatization. Various elements of these policies have created interest in the potential contributions of NGOs and have opened up more space for NGOs to participate in public-sector programmes. Seen from one angle, this implies more scope for NGO perspectives and capacities to be incorporated into state programmes. However, a converse view is that in the context of structural adjustment, NGOs will be able to have limited impact on public programmes. We will consider two particularly relevant aspects of structural adjustment programmes: the so-called social, or compensatory funds, and public-sector reforms.

Social funds and the NGOs

Opportunities for NGO participation are being opened up in many different public programmes, not just in the agricultural sector. One of the most influential experiences has been the participation of NGOs in 'social funds', above all the Social Emergency Fund (FSE) in Bolivia, which operated from 1986 to 1989 (Durán 1990; Newman *et al.* 1991; Wurgaft 1992). Deemed a positive experience, attempts have been made to replicate it in other Latin American countries – there are currently ten such funds operating in Latin America (World Bank 1990b; Sollis 1991; Wurgaft 1992). These funds are compensatory programmes to alleviate the 'social impacts of adjustment' on the poorest sectors of the rural and urban population. They attempt to channel social support, resources, small-scale infrastructure and employment generation projects towards groups that are particularly vulnerable to the unemployment, food price rises and cutbacks in social services that result from adjustment programmes. Some funds are set up on a temporary basis in the early years of adjustment; others, such as those in Bolivia, Chile and Peru, are permanent and based on the supposition that even successful adjustment will not improve living standards for the broad mass of people (Wurgaft 1992: 37).

Attempts have been made to implement some elements of these programmes through NGOs, who are asked to identify, design, submit for funding and then execute projects (Wurgaft 1992). Some of the reasons for this interest in NGOs stem from assumed strengths of NGOs. As multilateral and bilateral agencies begin to accept that NGOs have a more legitimate and effective relationship with popular organizations than does the state, and therefore a greater ability to reach the poorest sectors of the population, they have come to favour NGO implementation where possible (Ribe *et al.* 1990; World Bank 1991a).

On the other hand, much of the interest in NGOs stems from a desire to avoid channelling funds through the state. Given that one major objective of structural adjustment programmes is to reduce the size of the public sector, multilateral agencies have tried to avoid financing public-sector institutions with the social funds. Also, as the role of social funds is to 'buy time' for structural adjustment, alleviating supposedly short-term negative impacts that could otherwise stimulate popular unrest, agencies have sought means to spend the funds as quickly as possible. Given the state's notorious lethargy and inefficiency, they have looked to private agencies for their presumed greater efficiency.[5]

The experience of the social funds has been significant in a number of ways. It seems certain that it left the multilateral agencies, above all the World Bank, with a positive impression of NGO potential (Williams 1990). The agencies began to realize that NGOs had the capacity to implement projects, to be flexible, and to administer funds. These experiences with social funds have probably provoked an interest in promoting NGO participation in other projects.

The experience with the social funds has been equally significant for NGOs. It has been one of the first experiences of working closely and on a relatively large scale with the public-sector and multilateral agencies. For one of the first times the funds seemed to offer NGOs scope to influence the allocation of public resources. But it has also been an experience which for NGOs threw into sharp relief the potential contradictions of working with the state. While this contact has offered NGOs access to resources, and to discussions about the impact of adjustment, some fear that the fact that NGOs receive resources from the public sector could compromise their ability to criticize it. For this reason, some NGOs in Bolivia preferred not to seek FSE funds in order to maintain their independence.

NGOs also had the problem of deciding whether or not participation in the social funds meant endorsement of the structural adjustment programmes to which the funds were tied and which nearly all NGOs have criticized as socially regressive and unacceptable. In this vein, many commented that in implementing projects with social fund resources, NGOs would be providing nothing more than a social subsidy to the policies of adjustment: in short, they would end up being politically functional to the neo-liberal economic project. This may well be true – it may, however, not be that novel. By meeting the needs of sectors excluded from state programmes, NGOs had always in some sense helped dissipate popular protest. Indeed, they may have made it more likely that when things went wrong in these programmes the poor would criticize the NGO (who they knew) rather than the state. Commenting on the work of NGOs with the most marginalized sectors in Chile, Lehmann captures the irony of the situation: 'And as for Pinochet, he does not complain at all: have not the international agencies taken over his responsibility for the poor, and relieved him of the political pressure to change his policies? It is a most bizarre convergence of interests' (Lehmann 1990: 181).

Public-sector reforms in NARS: implications for NGOs

Amongst the various measures which make up the structural-adjustment programmes, the most important to our study has been the reduction of the public sector budget. These so-called public sector reforms, which are based on cutting the size of the state apparatus, also seek to tackle the problems of administrative inefficiency which arose as a consequence of the growing size, complexity and centralization of the public sector from the 1960s onwards, as discussed earlier (Martínez-Noguiera 1990; Selowsky 1990).

Some of these reforms merely rationalize public-sector activities, seeking to end situations where the functions of different public-sector institutions overlap. In other cases, however, the changes are more significant and some functions of rural development and ATD are being handed over to the private sector (Martínez-Noguiera 1990: 97). In this process of state withdrawal, it is important to distinguish between the privatization of the *implementation* of a given activity, and the privatization of its *funding*. In some cases, the state only privatizes the implementation of a service but maintains its funding responsibilities. In such cases, it is inappropriate to talk bluntly of 'privatization'. The purest case of this model is the provision of technical assistance in Chile, where the public Institute for Agricultural Development (INDAP) covers the majority of the technology transfer costs, but where private institutions carry out the work (see Chs 6 and 7). In Ecuador, the new National Rural Development Programme (PRONADER) proposes something similar, looking for NGO participation in the implementation of certain development activities that would be financed by funds channelled through the state (IICA 1991a; Bonifaz, personal communication, 1991).

However, when the state also hands over the responsibility for financing ATD activities to the private sector, the change is much more significant. In certain cases this change has not only a practical effect in the ATS (the public institutions cease to perform certain functions in agricultural development), but also an ideological one, which – according to some observers – is even more profound. They suggest that by ceasing to carry out certain activities, the state begins to question the principle that it has the responsibility to address the needs of the poorest sectors of society. In other words, by withdrawing from ATD the state may be contributing to a redefinition of the social contract between the state and different groups in civil society.

Cases do exist where we can speak of such a double privatization, of both implementation and financing. One recent example has been the change in IBTA in Bolivia. Previously, IBTA was responsible for agricultural extension, but it has now ceased to offer extension services in the Bolivian *altiplano* in the hope that NGOs and campesino organizations would assume a large part of this responsibility (see Chs 6 and 7). In a similar fashion, recent proposals in Peru for the restructuring of credit provision for campesinos recommended that NGOs take on responsibility for technical assistance, training and communication which were previously (in name, albeit not in practice) the responsibility of the public sector (IFAD 1991).

Another model implemented in a number of countries (such as Ecuador and Peru) has been the creation of a Foundation for the privatization of the funding, but not the implementation, of ATD. USAID has been at the forefront in these initiatives and has helped create five such foundations in Latin America and the Caribbean. Once established, the Foundation then receives funds from a donor and uses them to coordinate ATD and agricultural development activities, contracting services from institutions from the NGO, private and public sectors (Coutu and O'Donnell 1991). In the mid-term the goal is that these foundations should be able to raise private-sector finances for ATD (Pardey *et al.* 1991). In other cases a more partial process of privatization is being pursued, whereby the state progressively reduces its economic contribution to a given public institution until funding is cut off.

These reforms have opened numerous possibilities for a closer, more coordinated relationship between NGOs and public-sector institutions. In some cases, as will be analysed in Chapters 6 and 7, NGOs have attempted to take advantage of these possibilities.

However, it is important to recognize that it is not always clear that closer links with public-sector institutions are desirable for NGOs. There are various reasons for this. NGOs are not always convinced that the NARS are sufficiently strong to offer support to their work. Although it is true that public-sector institutions were already becoming weaker during the 1970s, the situation became critical in the second half of the 1980s. As CEPAL notes, '[w]ith very, very few exceptions, the public sectors of the region went into crisis during the decade. The excesses that had been committed in previous decades, reflected in bureaucracy, inefficiency and misallocation of resources, were painfully highlighted by the serious financial restrictions . . . in the 1980s' (CEPAL 1990: 35).[6] Once their budgets were cut back, the NARS, along with other institutions, became more inefficient and in some cases the majority of their budget was spent on maintaining their staff – an effect of their growth of previous years. Although some NARS programmes have managed to attract international funding, in general the period has been one of alarming institutional decline, less marked in Chile and Colombia, and more severe in other countries (Box 3.14).

Box 3.14 The crisis in IBTA

Throughout almost its entire existence, IBTA has experienced huge fluctuations in the size of its budget. While in some years the budget allowed it to contract qualified personnel and send some on to do post-graduate courses, in others the cyclical reductions in its funding meant that it had a high turnover of research and extension personnel. . . . In the last few years the situation has become so precarious that the majority of qualified professionals have left the institution and many of the extension agencies have been staffed by intermediate technicians or graduates.

(Bojanic 1991: 5–6)

This crisis has manifested itself in the NARS in various ways. The exodus of professionals which began in the 1970s has continued, many of them taking the training they received in the NARS over to the private sector, international organizations, and in some cases to NGOs (Piñeiro and Trigo 1983; Barsky 1990: 116). This exodus has been a response to the increasing scarcity of resources to maintain research and extension. In addition, as the inefficiency of already weakened institutions becomes worse, they often have even more difficulty in meeting any kind of commitment entered into with an NGO. Just as in the 1970s it was risky for NGOs to have contact with authoritarian governments, in the 1990s it is risky to enter into a relationship with public-sector institutions in fiscal crisis. Now, however, the risk for NGOs is not so much political as pragmatic (Carroll *et al.* 1991; CESA 1991). If the NARS does not fulfil its promise to work jointly with an NGO, it is the NGO which will suffer, as the quality of its relation with the campesinos with whom it planned to work is damaged.

Establishing relations with the state under these programmes of public reform is thus difficult for NGOs. They find themselves faced with 'a state which, in its restructuring, not only hands over to the private sector economic activities which were previously the remit of the state, but is also looking to civil society . . . for answers to social crises and even agricultural development' (Durán 1990: 27). NGOs now face the problem of whether or not, by accepting these new relations with the public sector, they will become its accomplices in the privatization of services which they had previously always insisted were a public-sector responsibility.

An additional cause for NGO circumspection is that to date the majority of these state proposals envisage a subordinate role for NGOs. Often it seems as if the state merely wants NGOs to implement state-designed policies in whose design and monitoring public sector institutions have shown little inclination to offer NGOs a say (Aguirre and Namdar-Irani 1991). In this sense, when public-sector proposals speak of NGO participation, what they mean is simply participation at the stage of implementation to improve or guarantee the efficiency of public-sector programmes; in very few cases are they talking about the democratization of these programmes (Bebbington 1991).

Finally, just as the unwillingness of the state to open up decision-making processes to civil society may not yet have changed, nor has its positivist and technocratic approach to agricultural development and research. The differences between NARS and NGOs understanding of technical change noted by Sotomayor seem as wide as ever, and remain untouched by democratization. Indeed, the gap has been aggravated by fiscal crises, for it appears that those who leave the state to go to international agencies tend to be those with post-graduate education who have been trained in both social and biological sciences.

Donors[7] and NGOs

At various points we have noted the important role played by donors in the development of relationships between NGOs and the public sector. They have also played a central role in the policies of structural adjustment and lately have shown interest in strengthening the democratization process (Healey and Robinson 1992). In both cases, they have identified roles for NGOs.[8] Also, now that the donors are showing more interest in the environment, they have been interested in the contribution environmental NGOs can offer their programmes. In the five countries of this study, multilateral agencies have been promoting a range of projects (agricultural and non-agricultural) with NGO participation.

In structural adjustment processes, donors (particularly the World Bank and Inter-American Development Bank) have identified various roles for NGOs. On the one hand, as we have noted, they see them as a means to implement programmes which the state cannot. On the other hand, they have demonstrated an interest in NGOs for their supposed capacity for flexible, honest and efficient administration. A further attraction of implementing programmes through NGOs is that NGOs do not imply a recurring cost in the public-sector budget. When an NGO programme comes to an end, it is the NGO which must look for funds to maintain its personnel. This stands in contrast to the situation in the public sector where the costs of personnel, offices and other institutional outlays continue after a project finishes.

Different donors have also been promoting NGO participation for more explicitly political ends: to strengthen civil society and the processes of social participation in the new democracies. During the last few years, these financing agencies have become increasingly explicit in acknowledging that democratization is both necessary for the development process and a desirable goal in and of itself (World Bank 1991a; Healey and Robinson 1992).

As argued in the governance debates in the World Bank, this democratization does not merely imply free elections, but also honest and accountable public institutions, local democracy and the participation of poor rural sectors in development projects (World Bank 1991b; EXTIE-World Bank 1990; ODI 1992). As such, democratization is enhanced by the grassroots organizing work done by NGOs (World Bank, 1991b) – a fact explicitly recognized in the new rural development programme in Ecuador (Box 3.15). Similarly, we have begun to impose a new 'political conditionality' on donors' loans – an insistence that the democratic processes must be respected in the borrowing countries. Although this 'political conditionality' has been applied on a wider scale in Africa (e.g. Malawi and Kenya), it has inspired approaches which have influenced donor policies in Latin America. This has contributed to an interest in designing programmes with NGO participation as a further means of exerting pressure over public-sector institutions, and demanding more efficient, honest and participatory practices on their part (Box 3.15).

These changes are significant for many reasons but one in particular merits comment. If donors are genuinely successful in forcing governments to be more responsive to demands from the poorer sectors of civil society, either directly or through their support to NGOs, they will have helped NGOs achieve one of their most important, long-held goals – the democratization of public administration and policy-making. Indeed, this donor concern for good government may make cooperation in public programmes between NGOs and

Box 3.15 PRONADER: an NGO role in development and democratization

PRONADER is an Ecuadorian programme of rural development supported and influenced by the World Bank and IICA. Its design gives NGOs a significant role in the administration of project funds, in the promotion of campesino participation, and in project implementation. NGO interest in PRONADER is due, in part, to its 'grand objectives which are to: a) Raise campesino incomes and improve living standards; b) Increase food production, and c) Strengthen the social fabric of democracy in the countryside by supporting campesino organization and participation' (IICA 1991a: 2).

Source: IICA 1991–2; Bonifaz 1991, pers. com.

donors who share similar goals a new and potentially powerful opportunity for NGOs to exercise influence over the state.

The problem of proliferation: eroding the legitimacy of NGOs?

Another effect of adjustment and democratization has been an extremely rapid growth in the number of NGOs in Latin America. This has been partly due to more secure political freedoms which have encouraged individuals to formalize and institutionalize their work with the rural poor. Nevertheless, the main reason has been the wider context of structural adjustment. Economic liberalization has aggravated the poverty of many people in the marginalized sectors of society, and concerned individuals have created NGOs in order to gain access to international funds and use them to respond in some way to the needs of the poor. In many regards, the ideology and motivation of these NGOs has much in common with the concerns of those of the 1960s and 1970s.

There is, however, a new and more opportunistic type of NGO that has been born of these changes. Whereas repressive governments created a politically displaced class of professionals, many of whom then created NGOs, structural-adjustment policies have created an economically displaced middle class, who have done the same.

Structural adjustment has hurt most professional middle classes. Some, particularly those in government, have lost their jobs as a result of public-sector cutbacks; others who have kept their jobs have seen their wages fall. As these people look for new or higher-paying jobs, the idea of creating an NGO has been increasingly attractive. Indeed, many NGOs can even pay wages in foreign currency, which are more likely to hold their value than wages in domestic currencies. As donor agencies' interest in working with NGOs is on the rise, so it has become easier to gain access to this foreign currency. Similarly, the mounting tendency of the state to sub-contract project implementation creates further demand for NGO services.

In response to this demand, new NGOs have been created, often by former government staff. In the case of Bolivia, Durán states that 'a large number of NGO employees . . . are technicians and intellectuals "relocated" from the state after the UDP government (1985), out of work, and with few prospects of employment given the recession and the chronic insufficiencies of the private sector' (Durán 1990: 91). Ultimately many such NGOs are more concerned about their income than about social change.

This surge in NGO formation is creating a number of problems. One is the immense difficulty of coordinating the actions of such a large number of institutions (see Ch. 5).

More significant is that public opinion has not been blind to what is happening, and one can sense a growing resentment of the ways in which NGOs emerge like mushrooms to gain access to these funds.

Although it is difficult to make hard-and-fast statements it would seem to us that the proliferation of NGOs is beginning to undermine their legitimacy and the very credibility of the label NGO. This creates obvious problems of credibility for the older, more radical NGOs. To call oneself an NGO by another name, as the Bolivian NGOs have done by adopting the label 'Private Institutions for Social Development' hardly solves this problem. They remain faced with the challenge that their legitimacy is being questioned by other groups in civil society – not least campesinos.

In the more specific context of relations with the public sector the surge in NGOs creates another problem. The 'opportunistic' NGOs have few qualms about accepting state contracts.[9] From governments' point of view it will be more attractive to work with NGOs who simply do the job defined in the contract rather than question the principles underlying the programme being sub-contracted. As donors channel increasing quantities of money through these contracts, such opportunistic NGOs are likely to grow, become more prominent and may begin to crowd out other NGOs. This may be the intention of donor agencies, though we doubt it. If it is not, however, they ought to think more carefully about how they go about working with NGOs.

CHAPTER SUMMARY

During the 1960s and 1970s, the period when NGOs began to emerge, relationships between NGOs and the state were generally tense. This was largely due to the political context in which NGOs emerged – they were a means of resisting and criticizing frequently authoritarian regimes. During the 1980s this context has slowly but surely changed in all five countries. Structural adjustment, donor pressures and decentralization are all generating more government interest in NGOs, and the return to democracy as well as the urgency of rural poverty challenges NGOs to rethink their relationship to the state. The increased contracting out of nominally public programmes has also led to the creation of a new breed of more opportunistic NGOs quite willing to accept government money and not especially concerned about sociopolitical change.

These changes have pushed the issue of NGO–state collaboration, in agricultural and other sectors, on to the agenda (Box 3.16). One message of this chapter is that exploiting these opportunities for collaboration will not be straightforward, neither will it always be possible or desirable. On the one hand, older NGOs remember the excesses of earlier regimes, and are concerned that the new democracies are still unconsolidated. Also the changes in context have not necessarily changed the bureaucracy and politicization in state procedures, and the positivist and technocratic approach of NARS to ATD. NGOs thus remain concerned to avoid the damaging effects such problems can have on NGO effectiveness. Furthermore, the fact that the state is looking for a way to shift some of its activities on to the private sector dissuades the older more radical NGOs from indulging such privatization, although it does not necessarily worry the more opportunistic types. Consequently, there is still much left to negotiate regarding the functioning of this relationship.

Box 3.16 Recent tendencies favouring increased coordination in the NGO–state relationship

- Political democratization and the weakening of authoritarian tendencies.
- Administrative decentralization in the public sector.
- Structural adjustment programmes and:
  - the reduction of state capacity: the state searches for 'counterparts',
  - positive experiences with NGO participation in the 'social funds'.
- Donor pressure on the state for it to work more closely with NGOs in search of:
  - efficient administration,
  - honest and accountable administration and good governance,
  - sustainable development.

Finally, a more coordinated relationship between the two sectors would imply a redefinition of their respective identities as well as their vision of their own institutional roles: NGOs, for instance, will have to reappraise what they understand by being a force for social and policy change if they begin to work with government; governments will have to reappraise their tendency to exclude NGOs and campesinos from decision-making.

None the less, one of the main conclusions of this chapter is that despite such problems the changes of the past decade have opened up more possibilities for NGO activities within their respective societies, and above all for constructive relationships with public-sector institutions. More than ever before, NGOs now have greater scope to influence the actions of the public sector, to scale up the effects of their programmes and to gain access to the human, technological and economic resources that reside in state institutions. If previously interactions between NGOs and the state have been conflictive, and linkages rare, the future likelihood is that linkages will be more frequent and effective than conflictive interactions. The rest of this book analyses the possibility and potential utility of exploiting these new opportunities, and some of the trade-offs involved.

NOTES

1 These are: the International Centre for the Improvement of Maize and Wheat (CIMMYT), founded in 1966, the International Potato Centre (CIP), founded in 1971, and the International Centre for Tropical Agriculture (CIAT), founded in 1967.
2 In his discussion of grassroots development in Latin America, Albert Hirschman (1984) discusses this phenomenon, calling it the 'conservation of social energy'. Hirschman observed that those who seek social change continue to reappear in different progressive institutions throughout their lives.
3 Lehmann notes a similar situation in the case of Brazil: 'even while CEBRAP [an NGO] was being branded as communist by some sections of the government, other branches were awarding them contracts' (Lehmann 1990: 184).
4 For broad indicators, see macroeconomic data in Annex 3.
5 Notwithstanding its significant channelling of funds through NGOs, it ought to be noted, however, that the FSE still spent only 29.4 per cent of its resources through NGOs (Durán 1990: 25).
6 In the Peruvian case, Torres speaks of 'the acute reduction in government capacity to act, expressed in the reduction of its programmes and in the fact that, due to the reduction in public sector salaries (to date, less than 25 per cent the purchasing power of the mid-1980s), the more experienced officials have pulled out' (Torres 1991: 4).
7 In this section we are talking about those multilateral and bilateral donors who work primarily with government.
8 Also, now that the donors are showing more interest in the environment, they have been interested in the contribution environmental NGOs can offer their programmes – a theme developed in the next chapter.
9 This is *not* to say that the NGOs that accept these contracts are all opportunistic; rather, if radical NGOs accept them, they do so as part of a strategy to promote social and institutional change. Opportunistic NGOs are mainly interested in being paid.

4

FROM MODERNIZATION TO A NEW TECHNOLOGICAL AGENDA

Campesinos and technological change in Latin America

In the previous chapter we argued that a series of political, administrative and macroeconomic changes have forced the issue of NGO–state linkages on to the agenda. In this chapter we argue that debates over the impact of Green Revolution technologies in the campesino sector are also leading to the conclusion that in order to adapt to current processes of economic and environmental change, future technology development strategies must shift their focus. They will have to move away from Green Revolution approaches based on centralized research institutes towards lower-input technologies based on a decentralized institutional structure. This will be necessary to meet both campesino needs and new challenges to the national economy. A lower-input and decentralized approach again suggests that NGOs and local organizations will have important roles to play in conjunction with NARS.

The chapter begins with a brief discussion of the introduction of Green Revolution technology in Latin America. It suggests that the ways in which new technologies were generated and incorporated were heavily influenced by factors of economic policy and sociopolitical power. This led to a pattern of technological change that was ultimately inappropriate for campesinos and Latin American economies alike. We then discuss two approaches to reorienting ATD: one based on scaling down Green Revolution technologies to fit campesino circumstances; the other based on agroecology. We suggest there is a clear common ground between the two which provides the basis of a new technological agenda in which NGOs and the state would have specific roles to play.

LARGE FARMERS AND CAPITAL-INTENSIVE TECHNOLOGY: BROAD PATTERNS OF TECHNICAL CHANGE

As we noted in the previous chapter, efforts to develop institutions for generating and transferring new agricultural technologies in South America took on momentum after the 1930s. This was part of the more general state-led initiative to dynamize Latin export-oriented economies reeling from the effects of the Depression in the northern hemisphere. This so-called Import Substitution Industrialization model of development aimed to break import dependency and envisaged the development of a domestic industrial base to serve national as well as export markets.

The roles assigned to the agricultural sector in this model were varied and to an extent contradictory, but in general terms the goal was to enhance domestic food production (to satisfy urban-industrial demands) and increase export competitiveness in those products with export potential. The introduction of new, science-based technology was to be the basis of this agricultural growth (Grindle 1986; Jennings 1988).

A second, subsequent basis to this growth were the land reforms of the 1960s and 1970s. Debates in the 1950s and 1960s argued that land reform should aim to redistribute land from the very large estates (*latifundios*) to the smaller farms (*minifundios*). Small farmers, it was argued, used land more productively than did the *latifundios* but their contributions to food production and economic development were constrained by the often semi-feudal ways in which they were tied to the estates. The conclusion was that these ties had to be broken and the market orientation of campesinos increased in order that through the incomes from marketing their produce they could then buy the products of domestic industries. There was, then, a strong economic argument for redistributing land, replacing a bimodal structure with a more unimodel structure (Johnston and Kilby 1975).

In the end, as a series of surveys of land reform suggest, relatively little redistribution of land in fact occurred (Table 4.1) (de Janvry 1981; Grindle 1986; Thiesenhusen 1989). What did happen, though, was that the threat of reform motivated many large landowners to incorporate technological innovations, modernize management practices and become more entrepreneurial (de Janvry 1981; Barsky 1984).

The continuation of a bimodal agrarian structure had important implications for both the sorts of technology that were generated and the ways in

Table 4.1 Areas affected by the agrarian reform and the number of campesino families benefited (selected countries)

Country	*Forest and agricultural surface affected (% of total area)*	*Number of farming families benefited (% of total)*
Bolivia	83.4	74.5
Chile	10.2	9.2
Colombia	9.6	4.2
Ecuador	9.0	10.4
Peru	39.3	30.4

Source: Prepared by the CEPAL/FAO Joint Agricultural Division, in Inter-American Development Bank, *Economic and Social Progress in Latin America, 1986 Report*, Washington: IADB, 1986, p. 130. Adapted from Thiesenhusen 1989: 10–11; de Janvry 1981: 206, for Colombia.

which they were incorporated. In essence, it had the effect of deepening a relatively capital-intensive pattern of technical innovation characterized by the incorporation of machinery, expansion of livestock activities, the use of agrochemicals and the installation of infrastructure for agro-industrial and export enterprise (Piñeiro and Llovet 1986).

Another of the linchpins of this technical change were the high-yield varieties from the international and national research centres, which, though often dependent on agrochemical inputs and irrigation, were in some cases widely adopted (Table 4.2). Use of inorganic fertilizers similarly proceeded apace (Table 4.3). Of course, these patterns varied among the countries considered here. This variation along with the exact factors causing this pattern of technical innovation are beyond the scope of this book (see Piñeiro and Trigo 1983; Thiesenhusen 1989b), but a couple of issues merit comment.

A strong argument can be made that a capital-intensive pattern of technical

Table 4.2 High-yield varieties of wheat and rice, by country (1970, 1983)

	% total wheat acreage 1970	*% total wheat acreage 1983*	*% total rice acreage 1970*	*% total rice acreage 1983*
Bolivia	1.9	9.2		
Colombia	9.2	95.0	17.4	91.8
Chile	61.2	70.0		
Ecuador	–	36.4	10.5	53.1
Peru			12.8	74.1
Latin America	10.8	82.5	–	33.7

Source: Adapted from CGIAR 1985, quoted in IDB 1986: 111, 113.

Table 4.3 Average annual fertilizer consumption per hectare of arable land and land under permanent crops

	1962–6	*1967–71*	*1972–6 (kg/ha.)*	*1977–81*	*1982–6*
Bolivia	0.8	1.3	1.5	1.3	1.7
Chile	24.9	29.5	28.7	21.7	31.1
Colombia	26.6	30.2	46.0	57.4	67.5
Ecuador	6.1	16.3	19.3	30.3	31.5
Peru	35.7	28.8	37.2	37.0	24.0
LA/Carib	10.4	18.0	28.5	39.1	40.1
Central America	21.1	34.9	49.9	54.9	53.7

Source: Adapted from FAO 1988: 25.

change was inappropriate for the rural and national economies of these countries. In contexts of relative capital scarcity it might have been more efficient to incorporate technologies that absorbed labour, given the lack of productive employment in urban areas, the steady increase of rural–urban migration and the spread of squatter settlements.

Many modernizing large farms chose machinery instead of labour, however. The reasons for this are several. One was a preference to avoid problems of labour management (Barsky 1984). More important though was a policy context in which exchange rates were overvalued and credit subsidized, reducing the cost to farmers of imported and capital-intensive technology.[1] Similarly, NARS tended to concentrate research resources in the generation of costly, often agroexport, technologies (Merrill-Sands and Kaimowitz 1990).

Macroeconomic and research policy thus favoured large farmers. This was not casual coincidence. Large farmers were able to exercise influence over the state and its policies (Piñeiro and Trigo 1983; Thiesenhusen 1989; Goodman and Redclift 1991a). Meanwhile, campesinos and their organizations were either too weak or too concerned with tenure issues to exert any challenge to this political power. They were therefore unable to demand policy and institutional arrangements favouring different forms of innovation (Piñeiro and Trigo 1983). Consequently, other technical problems that were primarily small-farm problems, such as the intensification of hillside production systems or the production of organic matter on small units, remained unaddressed by research institutions that responded primarily to large farmer demands (Uquillas 1992; de Janvry and Garcia 1992).

THE NEW TECHNOLOGIES AND THE RURAL POOR

Divergent interpretations

The debate as to whether the rural poor were able to benefit from technical change has been confusing and often acrimonious. At one extreme has been the argument that the campesino was excluded from the benefits of this modernization (Griffin 1975; Hewitt 1969). This exclusion was deemed to be partly an effect of the ways in which technologies were generated. NARS tended to conduct crop research on better quality land of the type that large farmers enjoyed but to which campesinos had less access; the inevitable result was that technologies generated were more likely to be appropriate to such large farmers and not to the complex and difficult conditions of campesinos. Likewise the researchers' focus on adapting high external-input technologies meant the technologies released were costly in cash terms. For resource-poor farmers these technologies were therefore either prohibitively expensive or at best, high risk because of the costs involved (Griffin 1975). Finally, the focus on technologies that replaced labour with capital was of less relevance to resource-poor farmers, and was particularly inappropriate for landless labourers who in effect lost out twice from this research focus (Lipton with Longhurst 1989): as they did not possess land, they could not make use of new technologies;[2] and as labourers, it was their jobs that were displaced when labour saving technologies were adopted by larger farmers (Thiesenhusen 1989).[3]

Not only were technologies inappropriate for campesinos but, it is argued in this critique of the Green Revolution, campesinos were excluded from access to credit, information, technical support and other services that would have helped them use and adopt these new inputs. The result, it was claimed, was to intensify social differentiation and the concentration of wealth and resources in the countryside. More recent studies continue to argue this position (Echenique and Rolando 1989). In many respects, the critique of agricultural modernization that claims that campesinos have simply not been interested in the new technology, and stick instead to their own superior indigenous knowledge, is also a replay of this argument.

None the less, in the years following these more critical interpretations, other empirical evidence emerged to suggest that campesinos were neither totally resistant to, nor excluded from, the benefits of this process of technical change. A body of research has argued that, while the new technologies did initially favour larger farmers, over time more of the benefits have been captured by poorer groups (Rigg 1989).[4] Increasingly it has been argued that distribution biases that emerged did not reflect problems with the technology *per se*, but rather with local institutional structures that led to social biases in access to the technology and to the related services that facilitated its adoption (Barsky 1990: 90; Rigg 1989). Indeed, some have gone so far as to argue that the rural poor have been able to use agrochemicals and new varieties as a means of trying to survive the poverty-intensifying effects of wider forces of commercialization and land subdivision (Bebbington 1992; Field 1990; Lehmann 1984; Rigg 1989).

This more positive reappraisal of the much-heralded ills of the Green Revolution has drawn increasing attention to the ways in which farmers have themselves adapted these technologies to make them more campesino-friendly. One important body of such evidence came from the 'Mantaro Valley' studies conducted in the Peruvian Andes by the International Potato Centre in the latter 1970s. These studies generated interest in the idea of farmer participatory research in CIP and its counterpart NARS. They showed that poor farmers experimented with, adapted and benefited from modern technologies within the limits of the initial technical packet (Horton *et al.* 1978, 1979). The implication was not that technological modernization was entirely inappropriate to such campesinos, but rather that a more participatory approach to generating modern technologies would lead to new 'packets' that farmers could put to better use.

The evidence for the experimental dynamism of campesinos is complemented by other research showing cases which suggest that in certain circumstances the agrochemical/modern variety packet has been the basis of a process of accumulation in which the poorer campesino households have been able to participate either directly or via sharecropping arrangements (Piñeiro and Llovet 1986; Barsky and Llovet 1983).

Reconciling divergence: closing the technology gap

We are faced, then, with what could appear to be contradictory evidence regarding the impacts of Green Revolution technologies in the small-farm sector. Some research suggests that the campesino was excluded from the benefits of new technologies. Yet other studies suggest that campesinos are engaged in a dynamic incorporation of modern technologies which leads to significant increases in yield *and* income.

The results, need not, however, be contradictory. Instead, they draw attention to the diversity of agroecological and socioeconomic contexts, and individual abilities and aspirations among campesinos and hence the differences in their response to modernization. Similarly, the studies point to the fact that just as there may be yield and technology differences between large and small farmers, so too there can be differences *among* small farmers. Indeed, such differences can be so great that some campesino practices are more similar to large farmers' than they are to those of other small farmers. Obviously our conception of campesinos must incorporate considerations of differences in local context, and differences among campesinos (Guggenheim 1989).

Evidence of the differentiated incorporation of modern technologies within the campesino sector, has inspired a number of studies aimed at understanding the causes of this difference. These have made recommendations regarding how to overcome the differences – that is, how to close the 'technology gap' (Cotlear 1989, Figueroa and Bolliger 1985; Joint Ecuadorian N.C. State University Subcommittee 1987; Echenique and Rolando 1989). In general, these studies have reached similar conclusions to the Mantaro valley studies at CIP – that is, that the non-use, or limited use, of modern technologies among campesinos is not so much due to the technology *per se*, but rather to contextual factors. This observation has in turn led to arguments that such contextual factors can be addressed, with the result that the process of technology diffusion could be extended more widely in the campesino sector. This, indeed, is one of the bases to Figueroa's proposal for a *via campesina* to agricultural development.

In many areas, such contextual obstacles no longer necessarily stem primarily from the power of landed rural elites. Other factors that have been identified as important in the incorporation of modern technologies in campesino agriculture have included the following:

- The nature and presence of technical assistance services. In trying to explain the wide yield gaps in Chile in the 1980s, Echenique and Rolando place much blame on the insufficient level of support of public programmes which reached less than 10 per cent of small producers in the late 1980s (Echenique and Rolando 1989: 57). These services must also make active attempts to adapt technologies to local circumstances, and not simply to transfer messages.

- The style of technical assistance. Appropriate technological support refers as much to how technology is transferred as it does to the technology itself (Jordan *et al.* 1989: 276). Orthodox extension methods and messages have often been criticized as inappropriate, but these general criticisms have been given much more precision by recent comparative research on the relationship between education and farm productivity in Latin America (Figueroa and Bolliger 1985). In a case study comparing three areas of campesino farming in Peru, each with similar agroecological characteristics but different degrees of modernisation (Cotlear 1989a, 1989b), Cotlear showed that the extension messages of state services were most appropriate and had most impact on productivity in areas of intermediate levels of modernization: for traditional areas they were too complex, and for advanced areas they were redundant. Moreover, they tended to become quickly obsolete: once the message was transferred farmers needed different types of support. In traditional areas, the most important impact on productivity came from primary-level formal education (giving basic numeracy and literacy skills). In modern areas most important was advanced formal education giving farmers more general skills of abstraction: Schultz's 'ability to deal with disequilibrium' (Schultz 1975). A large consultation in Ecuador also concluded that educational services were important in expanding the spread of technology (Joint Ecuadorian N.C. State University Subcommittee 1987). Popular education seems to be particularly relevant in traditional areas as a first step towards increasing literacy and numeracy skills, and heightening campesino awareness of their needs for, and rights to demand, more formal training in, among other things, the use of modern techniques (Carroll 1992: 60–2);
- The availability of credit. Campesino non-use of credit is often due to the inappropriate means by which it is administered, for example, the demand for land guarantees, bureaucratic slowness, excessively large quantities. It is not necessarily because campesinos do not want credit (Jordan *et al.* 1989; Barsky 1990). In cases where they have had easy access to credit, campesinos have engaged in rapid technical change (Barsky and Llovet 1986: 273–4).
- Access to dynamic and growing markets. Dynamic and growing demand for campesino production was one of the most important contributing factors in the cases of campesino innovation reported in Piñeiro and Llovet (1986: 343).
- Forms of local organization and market integration that guarantee that the

> campesino will be able to capture some of the benefits of technological modernization.

The impact of these contextual factors on the sub-system of technology use depends, in many respects, on macroeconomic policies, although the nature of the impact will vary between different localities. This reminds us that closing the technology gap will, in part, depend on macroeconomic decisions (e.g. resource allocation to education, credit and monetary policy, market development and regional integration). However, addressing some of these factors will also require the action and the support of local institutions helping, for instance, to adapt education style and technical assistance to local circumstances.

What unites the authors cited in this section is the belief that such contextual factors *can* be addressed, and that if they are, the benefits of technological modernization could be spread more widely and more equitably. Indeed, to the FAO such a strategy should lie at the heart of agricultural development in Latin America:

> A central requirement for achieving better performance in the future as well as stabilising rural populations and reducing poverty, is that the large population of smallholders and landless labourers must have better access to the technology, inputs and services they need to raise output and productivity. This means firstly recognising the untapped productive potential of these farmers; and secondly, making an active commitment to help them realise this potential through favourable public policies and the provision of greater public resources.
>
> (FAO 1988: 16)

However, the capacity of the public sector to address this challenge is by no means certain. Its past performance, scarce resources and orientation towards resource-rich farmers suggests other actors will also have a role to play: to pressure the public sector to respond to campesinos' needs, and to provide additional forms of support at a local level to NARS and campesino alike.

AN ALTERNATIVE TECHNOLOGICAL AGENDA: QUESTIONING THE GREEN REVOLUTION

The studies discussed in the previous section are based on the belief that it is more viable to work with the modern technologies that are already available in rural areas than to pursue alternative technological agendas. Barsky speaks for many when he observes:

> Attempts to develop so-called appropriate technologies at the margin of the overwhelming integration of campesino producers into existing markets have proven to be largely ineffective, whether pursued through public or private agencies . . . a [more] fruitful path is to implement . . .

> technological proposals for small farmers that make the most of the situations in which, like it or not, they currently find themselves
> (Barsky 1990: 89, 92)

Others are less sure. As the 1980s proceeded, the viability of the Green Revolution technical agenda has been questioned with increasing cogency. Questions have been asked of it on ecological, economic and social grounds.

Agroecology, a perspective rooted in ecology on agricultural development and resource management, has offered one of the most significant of these critiques of technological modernization. It is a perspective that has gained increasing ground in Latin America, particularly in the NGO community. Indeed, in many respects it has been built on Latin American NGO experiences (Altieri 1990).[5] Agroecologists have argued that sustainable production systems must be based on ecological principles and in, for instance, mechanisms of fertility maintenance and pest population regulation that occur naturally in ecosystems. Fertilization practices ought therefore be grounded in organic matter recycling, within systems that combine crop, livestock and trees; pests ought to be controlled by manipulating predator–prey relationships and natural pest-repellants. Typically, agroecologists look to indigenous, pre-modern knowledge systems as a store of popular wisdom regarding such practices.

The qualities of these practices are counterposed to the detrimental effects of agrochemical, fossil fuel-based technologies. Use of chemical fertilizers and pesticides, for instance, has, it is argued, damaging effects on soil structure, causes groundwater and surface water pollution, and creates dependencies on external economic and supply systems. Similarly, modern varieties and introduced crops are criticized for deepening dependencies on these inputs (an effect of breeding processes in which agrochemical inputs are used) and not exhibiting the qualities of local adaptation and resilience to a broad spectrum of local shocks that native crops and varieties typically possess.

Such technical and ecological constraints to the sustainability of the modern option are aggravated by economic constraints which the crisis and adjustment programmes of the 1980s have made more apparent. Agroecologists have long pointed to the risks to small farmers implied by the monetary cost of agrochemicals. These risks have become more apparent as adjustment programmes have led to removal of some of the price distorting mechanisms that had kept nominal prices for agrochemicals and imported machinery below real prices *and* that at the same time had subsidized the credit used (more by larger farmers than small) to purchase these modern technologies. As this double subsidy is slowly removed, so the real cost of the agrochemical and mechanized package has become more apparent (de Janvry and Garcia 1992). This means that even if the continued power of commercial agricultural interests means that other price distortions are maintained and that therefore 'the agroecological alternative has yet to be given a fair market test' (ibid.: 32), this situation is beginning to change. Increasingly the Green Revolution packet will have to be assessed against agroecology without the hidden support of favourable price distortions.

Along with such economic changes there has been mounting evidence of the effects of the environmental consequences of the modern technological packet. Forest clearance, forest soil degradation, pasture degradation,[6] hillslope erosion, water pollution, biodiversity loss and global climate change are environmental changes that inspire doubts about the sustainability of systems based on modern technologies. At the very least they question the sustainability of systems that do not incorporate indigenous and agroecological practices in combination with modern technologies (Browder 1989).

These empirical changes have been combined with ideological ones,[7] such as activities before and after UNCED, and the greening of a whole series of social and political debates (Cleary 1991). The combined effect is that, having been an alternative,[8] and thus a largely marginal and countercultural approach to agricultural development, agroecology and sustainable development have been thrust into an increasingly visible position. As we will see, Latin America NARs are not yet especially agroecological, but among donor and international organizations operating in Latin America one can see a realignment occurring.

Thus, some of the largest NGO and grassroots group donors in Latin America, such as the Inter-American Foundation (USA) and the International Development Research Centre (Canada) have reoriented their attention to sustainable production systems and agroecology. Perhaps more significant, though, are the changes in more orthodox institutions, such as the Research Councils, the International Agricultural Research Centres and the multilateral agencies.

CIAT, for instance, is currently in the process of shifting the organization of its research away from commodity lines towards eco-regional and land-use systems, reflecting an institutional effort to build sustainability and ecological concerns into its future ATD efforts (CIAT 1991). The focus of this ecosystem work will be Latin America. While CIP retains a commodity focus, it has initiated a large research programme in northern Ecuador into the environmental, health and economic implications of agrochemical input-intensive potato production (Crissman, pers. com.).[9] ISNAR, whose influence over Latin American NARs is exercised through research and consultancy visits providing support in institutional organization, research planning and management, has also recognized the limits to the Green Revolution package, and the growing need to reorient research agendas towards environment and resource management concerns:

> Almost all the increases in farm output required to meet global needs . . . must come from further intensification of production practices on existing agricultural lands. . . . But productivity gains from conventional sources are likely to come in smaller increments than in the past and to

> an increasing degree. These gains are likely to be crop-, animal-, and location-specific. . . . These concerns mean that research must place greater emphasis on developing new farming methods and land uses that act to sustain or enhance the natural resource base for agriculture.
>
> (ISNAR 1991: 21, 23)

Evidently, then, the terminology of so-called populist critics of agricultural research who spoke of low input, alternative and sustainable agriculture has now entered the lexicon of those they once criticized. Whether or not such claims of conversion to the agroecological cause are mainly designed to capture the initiative from the populist critics of institutionalized agricultural research (Ruttan 1991: 401),[10] rather than being genuine commitments to sustainable development, they are nevertheless significant for the influences one can expect them to have in individual Latin American countries.

According to several authors, the combined effect of these economic and ecological changes, and the realignments in institutional agendas, has been to elicit the emergence of an alternative technological agenda in Latin America (Kaimowitz 1991; Altieri and Yurjevic 1991). This agenda is based on sustainability concerns, the realization that it is no longer possible to maintain agricultural growth with capital inputs and price-distorting policies. It is also in some cases stimulated by the increasing urgency of reducing production costs in order to keep agriculture competitive in contexts of trade liberalization (FAO 1988: 16; IICA 1991; Kaimowitz 1991).[11]

This agenda will be influenced by perspectives from agroecology. A focus on soil management using labour, land transformation techniques and organic material will be central, as will work on diversified farming systems based on natural ecosystem processes. This in turn will mean agronomic research will become more important. Technological solutions will be location specific and so research will also have to be location specific (Ruttan 1991). Location specific adaptations in turn mean technologies will be information intensive rather than capital intensive. This will imply using more farmer knowledge (i.e. participatory ATD) but also providing support to farmers to increase their own knowledge and management skills (Kaimowitz 1991).

The emphasis of this agenda clearly differs from those proposals concerned to spread the benefits of modernization to the resource poor. How far it differs, though, depends on how pure an agroecological or modernizing line one seeks to take. Indeed, there is reason to be sceptical about any such purism. Resource poor farmers might understandably be aggrieved to see the rules of the technological game changed just as they begin to have greater access to new technologies (Bebbington, 1992). Furthermore, the problem of declining returns to inputs is more serious among high input commercial farmers than among campesinos, and in developed than in developing countries (Ruttan 1991).

AN EMERGING CONSENSUS ON A NEW TECHNOLOGICAL AGENDA: INSTITUTIONAL IMPLICATIONS

It is not our task to support one or the other of these two perspectives of closing the technology gap or pursuing agroecological alternatives. Indeed, it seems more useful to recognize the considerable overlaps between their messages. In their analysis of processes of technology change, to date both share the opinion that these have been biased in favour of larger farmers and have been based on research that has generated technological recommendations that are in large measure inappropriate to small farmers' realities. Similarly they both perceive ATD as having been insufficiently participatory.

The two perspectives also converge around a set of common themes that future technological agendas should incorporate:

- the generation of technologies more adapted to the types of agroecological conditions in which small farmers produce;
- the strengthening of institutional capacity at a local level in order to enhance location specific ATD;
- a widening of campesino access to support services that facilitate use of technologies, and more equitable access to those supports;
- a reduction of agricultural production costs, to increase on-farm income and food security;
- wider availability of information among campesino producers, and an enhancement of their capacity to use it;
- the conservation of the resource base in the long term, within the constraints of ensuring acceptable income flows in the short term;
- building an agricultural development strategy on the dynamism and knowledge of campesino producers.

Taken together these perspectives are the basis of a 'New Technological Agenda' (Kaimowitz 1991) that poses the challenge of closing the technology gap within the limits of the new economic and ecological context of post-structural adjustment and post-Brundtland Latin America. This is still the challenge of Figueroa's *via campesina* to base agricultural development on the enhancement of productive potential in the campesino sector.

CHAPTER SUMMARY

In this chapter we have argued that patterns of technical change in the post-war period in Latin America have been largely capital intensive and unsuited to campesino production conditions. They were also inappropriate for the region as a whole, absorbing a scarce factor of production (capital) rather than a more abundant factor (labour). This occurred as a result of economic policies that overvalued exchange rates and subsidized credit, and agricultural research policies that concentrated on capital-intensive production for favoured environments. This policy bias was in no small measure due to the political and economic influence of large farmer groups, and the inability of campesinos to exert effective pressure over policy-makers.

Although many critics argued that campesinos were excluded from the benefits of this process, there is considerable evidence to suggest that they are able to make use of these technologies when there are appropriate local support institutions which facilitate adaptation and use of these technologies, and which lower (but do not exclude) the use of external inputs.

The Green Revolution in Latin America has also been criticized on the grounds that it is ecologically and economically unsustainable. These criticisms are beginning to influence the agendas of many donor agencies and agricultural research institutions which show a clear shift towards an interest in lower input and locally adapted technologies.

Two lines of critique have thus led to similar proposals for new agendas for ATD. How far the state or the market will be able to respond to this agenda is questionable. Such doubts about the market and the state have drawn attention to the possible role of other not-for-profit, socially motivated local institutions. It is assumed that NGOs have a range of qualities that will allow them to (1) complement public-sector activities, (2) press that public sector to be more responsive to campesino needs, and (3) be important players in the new technological agenda (Altieri 1990; de Janvry and Garcia 1992; Kaimowitz 1991). Among other things, it has been suggested that NGOs are:

- more interested in low-input agriculture than the public sector;
- more able to adapt technologies to local conditions than the public sector;
- more participatory than the public sector, and so more able to link local and modern agricultural knowledge;
- more efficient in their own use of resources than the public sector;
- more oriented towards systems perspectives on ATD than the public sector;
- aware of and able to represent the aspirations of campesinos.

The truth of these suppositions about the relative capacities of NGOs and the public sector to contribute to this new technological agenda for ATD is the theme of the next chapter.

NOTES

1 In Bolivia, subsidized credit represented 78 per cent of public expenditure in agriculture in 1985. The Agrarian Bank of Bolivia channelled the bulk of it to the large agroexport sector of Santa Cruz. This cheap credit meant that the price of capital did not reflect its scarcity value (Morales 1990).

2 It should be noted that landless labourers may be able to use sharecropping contracts as a means of gaining access to land, and as a result be able to use these technologies (Lehmann 1982, 1984).

3 The exceptions were those cases where technical change in an agricultural sector was accompanied by overall growth of production (Piñeiro and Trigo 1983).
4 These groups may also have benefited as food consumers, by way of cheaper food (ISNAR 1992).
5 There is perhaps a correlation between the rise of agroecology and the crisis of left thinking: one form of critique has been replaced by another, and the object of critique has shifted from capitalism to modernity.
6 Although there is clear evidence of pasture degradation in Amazonia, there is still disagreement as to how widespread or inevitable this is (compare Hecht 1985, with Smith *et al.* 1991).
7 Such ideological changes, or changes in the discourse on development, were in part driven by the empirical changes, in part by a certain institutional opportunism (Ruttan 1991: 401).
8 In Altieri's original treatise he uses the word 'alternative' in its title (Altieri 1987).
9 This research, perhaps ironically, is concerned to evaluate the negative environmental impacts of the same campesino producers whose success in incorporating new technologies and initiating a process of capital accumulation had been celebrated by Barsky and Llovet (1983) and Lehmann (1984).
10 Ruttan was referring to the new found interest of the US National Research Council in alternative agriculture.
11 There is an interesting parallel here with the situation in the 1930s when competitive pressures were a factor in stimulating early state involvement in agricultural research (Pardey *et al.* 1991: 244–5).

5

INSTITUTIONS FOR THE NEW TECHNOLOGICAL AGENDA

NGOs and NARS

The agenda for ATD in the future as outlined in the previous chapters will demand an institutional structure that has the capability to:

- generate lower cost, environmentally benign technologies;
- adapt those technologies in different ways to different contexts;
- combine indigenous and external expertise through participatory practices;
- strengthen local capacity in campesino populations;
- sustain a process of continuous innovation to maintain competitiveness.

Institutional capacities will still be required in all the sub-systems of the ATS to meet this challenge. Basic and applied research will be required in order, for instance, to breed crop varieties that can produce in nutrient- and water-constrained environments, resist diverse pests and thus require less use of fertilizers and pesticides. Adaptive research and transfer activities will be required to combine external and farmer knowledge for locally adapted practices, and to disseminate knowledge of these practices.

In this chapter we consider in what ways the NARS and especially the NGOs will be able to respond to these challenges and in what ways they will be most constrained. We first focus on the capacities and weaknesses in the NARS that derive from their technological focus, resource base, organizational structure and approach to participation. Drawing on the case study material we then focus on similar themes for NGOs. The evidence suggest that acting alone neither NGOs nor NARS have the ability or resources to meet contemporary challenges. The future will require considerable changes in their respective approaches to ATD, and a far more carefully considered and open-minded approach to coordination among themselves.

THE NARS: TECHNOLOGICAL FOCUS, INSTITUTIONAL STRUCTURE AND APPROACHES TO PARTICIPATION

Technological focus

As we noted earlier, the approach to technology generation that was promoted at the Mexico Agricultural Programme and its successor, CIMMYT, and that underpinned the 'Green Revolution', was subsequently institutionalized in a continent-wide pattern of national research systems oriented to modernization, productivity increases and cheap urban food to sustain development models based on import substitution industrialization and urbanization. The cornerstone of such programmes has been the promotion of high-yielding varieties and the associated stimulation of agrochemical use (Box 5.1). Traditionally very little has been done by NARS in agroecology, or in work with native crop varieties and the recovery and improvement of indigenous practices (Altieri 1990).[1] To an extent this is now changing, although much more so in international centres and in parts of NARS where environmentally concerned donors are able to exercise leverage. CIAT in Bolivia is one such example of a NARS which has been increasingly concerned with agroecological issues, although largely as a consequence of leverage exercised by its donors. After a tradition of commodity-based research, it has initiated research in agroforestry systems and low-input options for small farmers in the Santa Cruz region, with the particular concerns of stabilizing slash-and-burn practices and limiting the damage incurred by mechanized farming.

Box 5.1 The NARS and technological modernization: INIAP and IBTA

The law creating INIAP in Ecuador defines as the institute's special function to collaborate with the Ministry of Agriculture in:

- organizing research centres to adapt and propagate seeds of HYVs, to improve crops native to Ecuador, and to select those that are most pest resistant and early maturing;
- soil analyses;
- studying and promoting the use of appropriate fertilizers, and in giving demonstrations of ecologically adapted cropping systems;
- introducing improved livestock and studying appropriate management practices;
- other research deemed appropriate.

Source: ISNAR 1989a: 13.

IBTA
'Research is in general restricted to testing genetic material that comes from international or Latin American organizations, or which comes directly from other countries' (ISNAR 1989b: 25).

Resources and institutional structure

None the less, the emphasis in NARS has been on crop breeding for yield increase. Such breeding required a concentration of public-sector resources in order to support basic and applied research with the necessary infrastructure of laboratories and a system of research stations located in different agroecological or administrative regions of the country. Each research station tends to have laboratory and library resources, though typically these facilities are strongest in the central research stations. These laboratories, greenhouses and research stations represent controlled environments in which 'ideal type' technologies are generated. Those technologies may then be tested off-station in regional trials to assess how they perform under different agroclimatic conditions. Finally they may be evaluated and adapted to farm-level conditions and then passed to transfer agents within the same institute or within a sister public technology-transfer organization.

This concentration of resources allows, theoretically, the NARS to operate in each of the sub-systems of the ATS at a national level. Thus in the Colombian NARS, ICA, for instance, basic and applied research are supposedly conducted in a national system of twenty-five experimental stations, while applied, adaptive and limited transfer activities occur in regional centres (Estrada 1991); in Ecuador INIAP conducts basic and applied research in seven experimental stations and ten small experimental farms (ISNAR 1989), with adaptive research and limited transfer work conducted in the production research programmes (PIPs); in Bolivia IBTA had twelve research stations but no real institutional mechanism for adaptive research, and over recent years an average total of 120 extension agents in eighty extension agencies for the transfer of research results (Bojanic 1991).

Nevertheless, it is questionable how far these NARS have been able to do basic and applied research, and even more questionable how far they will be able to do so in the future. The capacity to act in all sub-systems is a direct function of the resources available to the NARS. While resources were never abundant, they have been under pressure over the last decade (Table 5.1). This resource shortage has meant that the capacity to do research is increasingly constrained by mundane problems such as lack of chemicals in laboratories, and more serious problems of wage decline and loss of personnel capable of doing basic and applied research as they leave for higher-paying jobs in the private sector, international agencies and even NGOs. In Ecuador, for instance, INIAP's budget fell from US$2.9 million to US$1.3 million between 1987 and 1988, with a full 75 per cent of the 1988 budget going on personnel costs (ISNAR 1989a). In Bolivia, IBTA's budget has likewise fallen dramatically, and by 1988 was comparable to CIAT's, even though CIAT is responsible for just one department of the country, and IBTA for the remaining six (Fig. 5.1) (Bojanic 1991; Morales 1990: 26;). The years 1986, 1987 and 1988 saw the loss of 30 per cent, 28.2 per cent and 22.2 per cent of IBTA's professional staff, leaving only seventeen with MScs and no PhDs by 1989 (ISNAR 1989b).[2] Funding problems are not so critical in Chile nor in Colombia, where in 1992 ICA had a $66 million budget (Prager, personal communication, 1992). None the less, there too financial constraints are serious with increasing proportions of the budget going to staff and fixed costs (with proportionally less for research). As a result the tendency is increasingly for NARS to conduct less basic and applied research. They are becoming more dependent on international centres with the NARS screening and selecting genetic material coming from those centres. A second tendency is for NARS to conduct less transfer work and become more dependent in this sub-system on other national organizations (see Ch. 6).

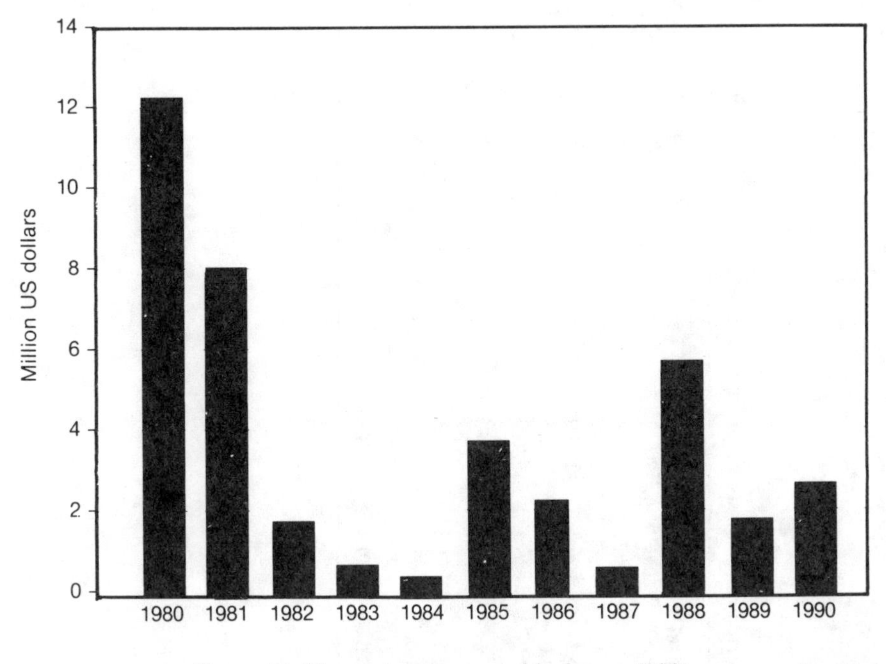

Figure 5.1 Fluctuations in annual budget at IBTA
Source: Bojanic, 1991.

Internally, NARS have been primarily organized along commodity lines. Efforts to introduce farming systems approaches have usually depended on donor funding. In the mid-1980s, for instance, ICA created a national farming system programme with IDRC support; when the support ended, so too did the programme (Estrada 1991). CIAT (Bolivia) also operates with a farming system focus, with bilateral ODA (UK) support.

This commodity focus has two additional shortcomings aside from the lack of a systems approach. First, the pressure of having to serve a national constituency has typically led to a proliferation of programmes: in 1988, for instance, INIAP was conducting research into seventy separate crops in twenty-three programmes and eight departments (ISNAR 1989). The inefficiency of such a strategy is clear: to spread scarce resources so thin renders any concept of achieving economies of scale in the NARS meaningless.[3] Second, socioeconomic research in the NARS remains weak or non-existent.

Table 5.1 Resources dedicated to public-sector agricultural research in Latin America

	Total agricultural research expenditures (millions 1980 PPP dollars per year)					*Total number of researchers (full-time equivalents)*				
	1961–5	*1966–70*	*1971–5*	*1976–80*	*1981–5*	*1961–5*	*1966–70*	*1971–5*	*1976–80*	*1981–5*
Central America	24.0	26.1	56.2	106.2	159.0	370	449	763	1262	1723
Bolivia	1.7	4.7	4.0	5.0	2.3	29	61	51	75	104
Chile	11.1	18.9	27.2	26.2	26.9	134	187	228	263	271
Colombia	33.3	46.7	40.5	35.7	47.8	326	398	559	400	454
Ecuador	3.4	6.2	16.2	17.2	13.3	34	54	138	180	211
Peru	6.1	20.7	19.4	11.6	20.3	121	174	217	256	262
South America	187.7	305.2	401.6	542.3	519.9	2014	3342	4713	5314	6774
Latin America and Caribbean	229.1	355.1	486.6	679.3	708.8	2666	4122	5840	6991	9000

Source: Adapted from Pardey *et al.* 1991: 418.

Consequently, issues of economic feasibility are imperfectly, if at all, factored *a priori* into the design of research programmes (Uquillas 1992).

Farmer participation in NARS

In a synthesis of the results of ISNAR's study of farmer participatory research in NARS, Biggs (1989) identified four different modes of participation in the different NARS reviewed:

- *contractual* - in which researchers merely hire inputs (land, labour) from the farmer, but make little effort to seek his opinions;
- *consultative* - in which the farmers' opinions are actively sought;
- *collaborative* - in which the farmer is in control of the particular trial or experiment;
- *collegial* - in which farmers and researchers interact as equals, the opinions of the former being taken on board by the latter when future research agendas are designed.

The four modes were of increasing intensity, from a simple contractual relationship to a collegial one in which hierarchies of status genuinely began to be broken down. In all four modes, though, a contact with resource-poor farmer production conditions, ideas and initiatives had allowed some degree of feedback influence over the NARS research.

Of the five countries we consider, there have likewise been experiences of formal participatory technology development with resource-poor farmers in all NARS except in Chile up to 1990 (though in INIA there *were* such relations with commercial farmers).

Case studies and interviews suggest these experiences have been primarily in the first of these modes, except for certain experiences of more egalitarian experiences which have depended more on the initiative and attitudes of individual researchers than on the NARS itself. Some, if limited, headway has therefore been made in campesino participation.

There are, however, two important observations on this. These participatory relationships are in the gift of the NARS (and their donors). Campesinos hardly ever have means of exercising formal accountability over the NARS, unlike large farmers who have at least had some form of representation in advisory and supervisory boards of NARS, and in some of the new Agricultural Research Foundations. There are isolated examples where campesinos have had some influence: in one of the on-farm research programme areas of INIAP in Ecuador a campesino organization has been able to exercise much influence over local NARS research, as a result of the general strength of the organization and its ability to police development activities in its area (Cardoso 1991). But these are isolated examples: the more general point is that participation has not led to mechanisms of representative democracy for campesinos within the NARS.

The second observation is that the NARS mandate is of course limited to ATD. Any campesino participation must, then, be limited to technology development. NARS cannot get involved in strengthening local organization, managerial skills and education. Similarly it is difficult for state extension services or the agrarian banks to allow flexibility in rules for credit provision unless endorsed in doing so by superiors, and often at a central level. The concept of farmer participation in NARS does not and cannot involve capacity building. As such, it is necessarily a quite limited conception.

Trends in the NARS: a summary note

It is of course difficult to generalize about the NARS in all the countries considered here: the NARS in Chile and Colombia are stronger than those in Peru, Ecuador and Bolivia, but INIAP in Ecuador has had a better developed conception of campesino participation than did Chile's INIA under Pinochet. Nevertheless, certain patterns and trends seem apparent and relevant to our concerns.

First, NARS have in general sustained a focus on Green Revolution types of technology and have only limited (though increasing) experience in low-input and indigenous farmer practices. Second, NARS do have infrastructure for actions in all sub-systems of the ATS, but their budgetary problems mean that the quality of those actions is deteriorating, and will be all the weaker as long as NARS try to spread resources thinly among all sub-systems. If NARS are to play a part in responding to emerging technological challenges they will have to concentrate and target their resources far more strategically than has been the case in the past.

Third, while NARS have addressed issues of participation, their mandate means that their conception of participatory development is necessarily limited. They cannot engage the wider issues of capacity building among campesino groups that authors like Carroll (1992) have shown is elemental in grassroots development, and others like Cotlear (1989a) demonstrate as vital to effective technology use by campesinos.

NGOs: TECHNOLOGICAL FOCUS AND INSTITUTIONAL STRUCTURE FOR ATD

Technological focus

In NGOs, a diversity of technological agendas have been pursued. These range from modernist perspectives that are indistinguishable from those of the NARS, through to radical agroecological positions. For ease of interpretation we have divided these perspectives into three:

- Modernist NGO: the NGO endorses the validity of a scaled down version of the modern technological packet.

- Agroecologist NGO: the NGO rejects modern technology and seeks to promote technologies grounded in the principles of ecology.
- Ethno-development NGO: the NGO rejects modern technology and seeks to promote technologies grounded in indigenous knowledge. While these may follow ecological principles, their validity lies in their being indigenous.

Notwithstanding these different technological perspectives, the NGOs in this study share a common sociopolitical agenda: the 'empowerment' of the rural poor (Lehmann 1990). We need then to look elsewhere to understand their differing technological agendas.

One factor explaining this diversity is that the NGOs work with campesinos farming in quite different circumstances. Some campesinos have incorporated modern techniques as part of their livelihood strategies, and in encountering this reality many NGOs have used it as a base from which to work: strengthening the strategies in place, rather than aiming to replace them (see Annex 2, Profile I, AGRARIA; Profile 2, CESA-Ecuador, and Profile 3, CIPCA; cf. Chambers 1987).

To some extent, the diversity of perspectives also reflects differences of opinion among NGOs regarding the nature of appropriate development. Thus, a modernist NGO may argue that it is more important for ethnic Indian groups to share in the fruits of national development and have access to the new technologies that have typically been the preserve of large and commercial farmers; that is, the modernists are concerned to close the technology gap discussed in Chapter 4. Ethno-developmentalists, on the other hand, suggest that in the face of the cultural homogenization that comes with Green Revolution strategies, ethnic groups should revalidate and use indigenous technologies as part of a broader effort to define a separate cultural identity. Agroecological NGOs would suggest that sustainable development has to be grounded in environmentally benign technology. These latter two types of NGOs are therefore more concerned to launch the radically different technological agendas discussed in Chapter 4.

This general debate over technology also reflects the general debates and uncertainty over the appropriateness of either modern or agroecological strategies. Among all the NGOs in the study there was growing concern about the rising cost of the modern technological option under current neo-liberal economic programmes, leading many to begin questioning the validity of modern technologies. Several NGOs are indeed moving from a modernizing focus to more agroecological concerns (see Annex 2, Profile 5, CESA-Bolivia, and Profile 6, FUNDAEC). On the other hand, many are worried that while agroecological strategies appear attractive, it is difficult to incorporate them into campesino systems without engendering a fall in production and income, at least initially (CESA 1991). If the NGO were to promote a technology that caused such a drop in income, its relationship with farmers would be damaged. Indeed, it is for this very reason that many NGOs initiate work with simple,

Box 5.2 Approaches to technology in NGOs

Among the NGOs selected for the case studies, there was a range of perspectives on technology. The table below summarizes examples of elements of the technologies they promoted, the client groups involved and the agroecological zone.

	NGO	*Client group*	*Agroecological zone*	*Technological focus*
Modern	CESA-Ec	Indian highland peasant	Rain-fed Andean slopes	Modern varieties, limited use of agro-chemicals, forestation and limited terracing.
		Lowland mestizos	Floodplain (the coastal plain)	
	AGRARIA	Mestizo campesinos	Semi-arid coastal brushlands; humid temperate lands	Modern varieties, fertilizers, vine cultivation, livestock vaccination.
		Mestizo labourers	Semi-arid coastal brushlands; humid temperate lands	
	CIPCA	Indian colonists	Lowland forest margins	Modern varieties, increased sowing density, machinery.
	SEPAS	Mestizo campesinos	Rain-fed, semi-arid northern plains	Introduction of new crops, irrigation techniques, agroindustry.
Agro-ecology	El Ceibo	Indian colonists	Mid-slope forest margins	Organic production of cacao and rural agroindustry.
	CESA	Mestizo colonists	Lowland forest margin	Agroforestry, legume winter crops and indigenous soil classification to manage slash/burn practices
	IDEAS	Highland Indian	Andes; rain-fed slopes and plains; semi-arid coast	Diversified low external input systems and kitchen gardens; rural agro-industry.
	CIED	Highland Indian	Andes; rain-fed slopes and plains	Diversified low-input and 'alternative' technologies; micro-drainage basin approaches.
	GIA	Mestizo campesinos	Rain-fed semi-arid and humid	Farming system development, some mechanization.
	FUNDAEC	Highland and valley mestizos	Rain-fed mid-altitude slopes	Alternative systems, IPM, resource conservation.

	NGO	*Client group*	*Agroecological zone*	*Technology focus*
Ethno-devel-opment	CAAP	Highland Indian	Andes; rain-fed slopes	Recovery and im-provement of traditional rotation systems, native crops and native trees

Note: The diversity of approaches among the NGOs is significant, for in many regards they share similar sociopolitical agenda. This reminds us in part that there are different opinions regarding the appropriate technological strategies for a particular political project; it also reminds us that, as they embark on pursuing that project, NGOs encounter great diversity in the production and living conditions of different groups of the rural poor: they do not all start from the same point. Their different technological agendas are thus also responses to different circumstances.

Source: Case studies.

high pay-off, Green Revolution techniques in order to gain farmers' trust (see Annex 2, Profile 1, AGRARIA). One of the few agroecological options that has not led to a fall in income has been El Ceibo's successful promotion of organic cocoa production for export to European health food markets. There were, however, special reasons for this success. El Ceibo enjoyed access to a high price market facilitated by European 'fair-trade' NGOs (Milz 1990; Trujillo 1991). It would never have been possible to achieve such prices in Bolivia, where as in the other countries considered here, organic and health-food markets remain very small.

Problems with the different technological options feed lively debates inside these NGOs. This is important to recognize, for while in Box 5.2 we have highlighted the 'main tendencies' of the NGOs, internal to each organization there are different currents of opinion. As a consequence one can sometimes find different projects with different technological foci *within the same NGO*. This is part of the NGOs' experimentation with new options, and part of their way of coping with internal differences of opinion.

What seems particularly important about this diversity is that it increases the possibility of innovation within and amongst NGOs. Such flexibility to experiment is aided by the NGOs' institutional structure and is far less frequently encountered in the more rigid structures of NARS.

Resources and institutional structure

Individual NGOs: size and management issues

In their rhetoric, NGOs argue that their institutional organization is the opposite to that of NARS. While concentrating resources in a NARS represents an effort to exploit economies of scale, a supposedly central principle behind NGOs' institutional organization has been to avoid the diseconomies of scale inherent in the bureaucratic rigidities and centralization of decision-making that are so often characteristic of large organizations.

There is without doubt some truth in this. However, NGOs also feel the tension between remaining small and the power and economies of scale that could come by growing. Some have therefore chosen to grow.

Of course, there are many NGOs who only have a handful of professional staff, with a few additional support staff. Such NGOs sustain field activities in one or two locations, on a small scale and with only a local presence. Among the NGOs selected in this study only CESA-Bolivia was of such a small size, with an annual budget in 1991 of around US$120,000 (Kopp, personal communication, 1991). As we will see, this has not stopped it from being innovative in the methodologies it has developed (see Annex 2, Profile of CESA–Bolivia).

Other NGOs, however, are far larger, and command budgets into the millions of dollars. CESA-Ecuador had in 1991 a professional staff of fifty-six with forty-seven support staff distributed among seven regional offices and eight regional projects. To support this, its annual budget was around US$1.7 million. AGRARIA in Chile similarly had over 100 staff in 1992, distributed among nine of its own projects and a further seven under contract from the state's technology transfer programme, INDAP (AGRARIA 1991; see Chs 6 and 7, below). An inter-NGO rural development programme in Bolivia coordinated by the Association of Popular Education Institutions (an NGO network), enjoys an annual budget of US$2 million, and so on (Bojanic 1991). These are the larger NGOs. Others, such as FUNDAEC, in the mid-range but well established, enjoy budgets of between US$0.5 and US$1 million.

Such budgets and staff numbers should not be compared directly with those of NARS. Only a small proportion of these sums are specifically for ATD, and much goes on other areas such as credit, training, health, and so on. Thus, of FUNDAEC's US$600,000 about US$350,000 was for ATD activities. Indeed, a more adequate comparison would be between total NGO budgets and those of state rural development programmes which likewise combine these different lines of action; on that comparison NGOs still appear small, when measured, for instance, against the budget of US$120 million for PRONADER, the new five-year programme of rural development in Ecuador.

None the less, it is clear that some NGOs have become large and quite powerful institutions. One advantage of this has been that they have been able to operate in the different regions of their countries. This allows them a chance to make more informed comparative assessments of agricultural development strategies. It also allows them to generate a range of lessons that are adapted to differing agroecological and regional conditions (Aguirre and Namdar-Irani 1992). Size permits the NGOs to generate overheads that allow resources and staff time to be devoted to activities not accounted for in particular projects (an economy of scale), such as staff training, discussions and negotiations with other institutions, and writing and research. It also allows the NGO to maintain specialists (e.g. in agriculture, gender issues and monitoring and evaluation) who can provide support in specific areas to field projects.

Growing in size, however, requires careful forethought if it is to generate benefits. Many NGOs do not plan for their growth. In a survey of twelve NGOs in Peru, Torres (1991) observed that many NGOs grow, but maintain the internal structure or the work procedures they had when they were small groups of professionals united around one or two key leaders. The result is that NGOs can become little more than confederations of projects, and fail to reorganize themselves along thematic lines. They fail to mature as institutions.

This decision to grow can have its own contradictions, though – as has been discussed in the scaling-up debate (Annis 1987; Edwards and Hulme 1992). On growing so large, NGOs begin to suffer some of the public-sector problems of centralization and bureaucratization from which they claim to be exempt. Local NGO staff in particular projects complain about the tendency of central office to interfere in their work, and to take excessive overheads from project funds. They also complain that key individuals control positions of authority in the central office, thus blocking any real possibility of career advancement within the NGO. These are real problems, with no easy solutions because of the trade-offs involved between staying small and cosy, or growing large and assertive. One NGO response has been to look for ways of increasing the autonomy of regional offices. In the most extreme case (AGRARIA), regional offices are to become legally autonomous parts of a sort of 'federated' NGO and will have the legal authority to raise their own funds.

This tendency toward bureaucratization and centralism in NGOs must be recognized, because much of the rhetoric about NGOs is that they are small, efficient and internally democratic. Furthermore, this rhetoric comes from those such as the World Bank who, on the grounds of this supposed efficiency, want to give NGOs more resources – that is, make them grow. Already NGO experience suggests that to do so may undermine the very efficiency that is being sought.

Centralism is not, however, only an effect of size. It can also be due to legal structures or institutional histories which vest much authority in key individuals in the NGO (cf. Carroll 1992). On the basis of this authority, those individuals are able to exercise inordinate personal influence over the NGO's actions. Indeed, during the course of the research, one of the NGOs studied was undergoing a process of radical restructuring and staff change that appeared to be directed by a few key people in whose decision-making field staff seemed not to have participated.[4]

These caveats are important. NGOs may claim to be efficient, internally democratic and flexible, but there are genuine managerial limitations that need to be addressed and which call some of these claims into question. None the less, once again the critique ought to be put in proper perspective. While we should assess NGOs' actions against what they claim to be able to do, we must not lose sight of the other criterion against which NGOs should be measured: the situation in the public sector. On this measure NGOs perform well. Characteristics of their institutional structure (such as relative smallness, relatively high levels of local autonomy in field programmes) allow them greater flexibility in responding to problems as they arise, and taking into account additional dimensions of the local food systems as they become relevant. Even when the central office is involved in local decisions, in NGOs this influence is generally less rigid, and is executed much more rapidly, than in the public sector. This allows field workers more opportunity to follow their own initiative and gives them more discretion over the use of resources (Box 5.3).

Box 5.3 Central office/local office relations: NGOs and NARS compared

A comparison from Ecuador illustrates some of the benefits of greater institutional flexibility in NGOs: that between the NGO CESA and the public sector's on-farm research programme, the Programa de Investigación en Producción (PIP). In both organizations the relationship between field staff and programme director is good, and the directors are most willing to support field workers. However, there are important differences. On the one hand, in CESA, central office senior staff are responsible for only three or four field projects and therefore can give them more frequent support than can the Director of the PIPs who must work alone in supporting nine different field teams. And second, to approve field staff's spending requests the Director of the PIPs must gain Ministry of Agriculture approval – which often arrives after the need for local expenses has passed. By contrast, in CESA authority can be granted much more quickly. Moreover, if PIP workers confront a problem that does not fall within the PIP mandate, they will not be permitted to spend money in response to it; conversely, CESA field staff have more flexibility to respond to different types of local need. Consequently CESA's field staff can respond more rapidly and effectively to local problems. The National Director of the PIPs was quite aware of this, and often compares these aspects of CESA's work favourably with his own programme, and would change accordingly if he had authority (Cardoso 1991).[a]

[a]This Director has since left INIAP.

The NGO sector: size and functions in the ATS

A small number of NGOs are predisposed to conducting research. A particular strength in this regard is socioeconomic research analysing the context of the ATS and its interrelationship with patterns of technology generation, use and impact at national and local levels. Indeed, some would argue that research NGOs have been the most important national source of analysis concerning the relationship between the context and functioning of the ATS in the countries reviewed. This strength has largely been a consequence of the weakness or repression of critical social science in the universities, which has led academics to join, or form, NGOs to do this research. Working in an NGO allowed greater freedom, often higher wages and the chance to link research to a particular project or policy focus.[5]

Far fewer NGOs do systematic agronomic research, and in general the limited concentration of resources in individual NGOs has meant that very few engage in applied research and none in basic research. Some NGOs, partly because they lack research stations, partly as a matter of principle, conduct this research on farmers' fields. Others have maintained research stations, though with less infrastructure than NARS stations. The motivation for conducting such research is generally to address issues untouched by the NARS and other research centres. Agroecological themes and indigenous knowledge have thus been a priority. In Ecuador, CAAP has conducted research on native crops and the soil fertility and pest control implications of different rotation systems based on farmer practices (see Box 5.4). In Chile, the Centro de Educación y Tecnología (CET) has conducted agroecological cropping systems research, and again in Ecuador the Fundación Brethren Unida (FBU) mounted a research programme to screen indigenous technologies (Zuquilanda 1988).

In undertaking such work, NGOs are engaging in a form of research that is extremely weak yet urgently needed in Latin America. As interest in the role of indigenous technologies for sustainable-resource management grows, a more systematic evaluation of the validity of such technologies must be conducted (Wilken 1989). The NARS are not doing this.

However, NGO capacity to do such research well is limited by a series of obstacles. Agroecological research generally requires a longer time-frame than donors are willing to support (CAAP therefore maintains its research station with its own funds); the data generated can be so complex that it requires analytical capacities NGOs do not possess (Box 4.4); and NGOs often lack the resources and facilities (e.g. laboratories) to conduct all analyses necessary. Finally, the action orientation of donors and NGOs, and the competitiveness among NGOs to show impact, is an obstacle to careful research and can lead to technologies being released before they are ready (Kohl, 1991).

Given such obstacles, some relatively well resourced NGOs argue that they ought to restrict themselves to the adaptive and transfer sub-systems, and not conduct basic and applied research (see Annex 2, Profile 3, CIPCA). Smaller NGOs do not have any choice. Consequently the bulk of NGO activity is confined to the adaptive and transfer sub-systems of the ATS.

When NGO activity in these sub-systems is aggregated the quantitative importance of NGOs as institutions in ATD becomes clear. Although data on the number of NGOs operating, and on their staff totals, is extremely difficult to obtain – partly because of the diverse ways in which they are defined and partly because there is little or no registration of NGOs in these countries – some estimates of the number of NGOs working in rural areas are presented in Table 5.2. When the number of field staff is summed, it appears that the gross contribution of NGOs can be of a similar order to that of the NARS. In Bolivia, Bojanic (1991) estimates that against the 120 transfer staff of IBTA, and the

Box 5.4 An NGO's agronomic research station: CAAP

According to CAAP's analysis of ATD in the Ecuadorian public sector, the technologies it generated were appropriate for a social, economic and bio-physical context that was different from that of the Andean indigenous peasantry. CAAP therefore aimed to initiate a process of technology generation that built on campesino production rationales. Field research identified the following as the main elements of this logic:

- long-term ecosystem conservation;
- food production for domestic, ritual and social needs;
- optimum use of available labour power.

These formed the basis of a programme of technology generation at CAAP. This began with research conducted on campesinos' fields, but researchers found this to be too difficult to control. CAAP therefore bought a parcel of land in typical campesino production conditions (rain-fed, sloping and with poor, eroded soils) which became its experimental station. The research programme concentrated on: (1) improving local technologies and Andean crops; (2) soil conservation; (3) fertilization trials; (4) pest control via rotations and traditional controls.

There were, however, many shortcomings in this early research. While senior staff had conceived it as a research programme, many field staff perceived it as an effort to use the farm to show local campesinos the benefits of terracing. It therefore became more an exercise in demonstration than research. Ironically, in working with terracing it also worked with introduced conservation techniques, for despite its rhetoric, CAAP promoted non-indigenous technologies, albeit agroecological ones. Looking more closely at farmer techniques and goals, it was decided that if research was to build on these it should concentrate on: (1) yield increase; (2) fertility maintenance; and (3) pest control. It did so with a programme of research on rotations and local varieties. Variety testing was linked to a collection programme.

The research has generated new information for Ecuador on the fertility effects of different rotations. In this sense CAAP contributes to filling a gap in ATD research in Ecuador with some success. Ultimately, though, the quality of its contribution is limited by constraints on institutional capacity: lacking expertise in trial design and evaluation, CAAP now finds itself with a mass of data that it is finding difficult to analyse.

Source: CAAP 1991; interviews.

300 that there are in all public agencies, there were 1,200 NGO staff working in technology transfer in 1991. Back in 1988, Gómez and Echenique (1988: 244) estimated that in Chile there were 754 professionals in NGOs directly linked to rural development, a figure similar to the number working in the public agency responsible for technology transfer and social development in rural areas.

The work of many NGOs in the adaptive and transfer sub-systems is particularly interesting because they combine ATD work with other actions aimed at tackling contextual factors that constrain campesinos' ability to use technologies. These other activities combine organizational strengthening, popular education, credit provision and off-farm activities such as product

Table 5.2 Estimates of number of NGOs working in rural development

	Source	*Number of NGOs working in rural development*	*Total NGOs*[a]
Bolivia	Bojanic (1991)	400	600–750
	FAO (1990)	153	
Chile	Gomez and Echenique (1988)	61	
Colombia	Prager (pers. comm., 1992)	*c.*25	*c.*50
Ecuador	Barsky (1990)	over 100	
Peru	Carroll *et al.* (1991)		350

[a]By NGO we refer to what were termed 'GSOs' in the Introduction, above.

processing and marketing. Such complementary interventions increase the likelihood that campesinos will be able to benefit from new technological options. The freedom to engage in such a breadth of activities is a direct effect of the institutional autonomy and flexibility that NGOs enjoy and NARS, by mandate, cannot. We consider the strengths, and weaknesses, that derive from these institutional qualities, perspectives and forms of intervention in the following section.

NGOs AND ATD: KEY STRENGTHS AND WEAKNESSES

NGO strengths

If the main strength and success of NARS has been to release technologies, the main strength of NGOs has been to elaborate methodologies that help disseminate and adapt technical information to the campesinos' local reality. Similarly, if many of the main weaknesses of NARS have stemmed from their institutional, political and economic context, many NGO strengths stem from this context.

In the following, we spend most time discussing the institutional and methodological strengths of NGOs. This reflects one of our conclusions from this study: that in the rush to link public-sector organizations and NGOs, discussions ought not to revolve only around the question of how to *combine* public-sector organizations and NGOs in order to improve the efficiency and effectiveness of ATD programmes. Far less should we be thinking of how far it is possible to devolve as many tasks as possible to NGOs. Of equal importance, we should be asking what it is that NGO experiences can teach us regarding the characteristics of an institutional structure that meets the challenge of making the ATS more democratic and more able to generate appropriate technologies – and how far can these lessons be used in restructuring the NARS and thinking about their relationships to NGOs.

Systems approaches and wider perspectives

Most of the NGOs in this study, and most of the strongest ones in the region, use systems approaches to their work. Nevertheless, there is considerable diversity among them on how to conceptualize a farming system. Even within NGOs, there are different approaches to the concept, inspiring both debate and internal critique. In the Chilean NGO, the Grupo de Investigaciones Agrarias (GIA), there have, for instance, been two competing concepts of farming system (Sotomayor 1991). The one more in the vein of certain tendencies in Anglo-American Farming Systems Research is strongly quantitative and model-based. The other is based on a more descriptive analysis of farming systems, tracing its roots to a French tradition that, in part because of the post 1973 diaspora, has had considerable influence among Chileans who have subsequently returned to work in rural development. Adherents of the latter perspective have criticized the former for spending unnecessary resources collecting and modelling data into systems models that in the end are far too complex to be directly useful for GIA's development work.

This debate on systems concepts within NGOs is indicative of their greater propensity to intervene in different parts of the campesino economy to facilitate the incorporation of new technologies. Indeed, many NGOs go beyond a farming systems perspective to embrace *food systems approaches* which combine on-farm and off-farm concerns in their work. Among several Chilean NGOs, for instance, this combination is reflected in their attempts to combine ideas of production systems and what some of them call 'micro-regions' (e.g. Sotomayor 1991). The concept of a micro-region integrates on- and off-farm themes, in the context of regional ecological and socioeconomic conditions. In this regard the idea bears some resemblance to the idea of a food system. Among Peruvian NGOs the concept of a micro-drainage basin (*microcuenca*) has similarly been very important, both in their attempts to incorporate environmental issues into their thinking and in the design of projects which have sought to use such physiographic units rather than administrative units as their spatial building blocks. In all these formulations, what one sees is a desire to combine crop and livestock production and the environmental context with other issues, such as product transformation, marketing and credit.

The potential relevance of combining on- and off-farm concerns is that, by increasing on-farm rates of cash accumulation, the propensity and ability of farmers to adopt technical improvements could be enhanced – because of increased monetary income and subsequent reduced risks involved in incorporating new technology. In Ecuador's province of Cotopaxi, CESA has for several years worked with an Indian federation in the improvement of barley production technologies. Recently, encouraged by the federation itself, CESA has sought outside assistance in the installation of a campesino controlled barley milling plant, to process the greater quantity of higher-quality barley now being produced. In Salinas, also in Ecuador, FEPP has worked with the local federation of communities combining cheese production and marketing with credit and technical assistance to help with the improvement of livestock and pastures. Subsequently, along with other NGOs, it has begun to work with the federation in forest management technologies to respond to overgrazing and deforestation pressures that were the result of the economic success of cheese production which induced farmers to expand their pastures (Bebbington *et al.* 1992).

In the Alto Beni of Bolivia, El Ceibo, a campesino federation, began with the goal of improving the marketing channels for campesino cocoa producers. It has subsequently combined these activities with a technology screening, adaptation and extension programme for the generation of organic cocoa production technologies specifically oriented to a European market for health foods (Trujillo 1991). This is a clear case of a membership organization successfully developing production technologies for a specific food system context – with the net effect that campesinos enjoyed higher prices for their product (Healey 1988; Tendler *et al.* 1988; see Annex 2, case study Profile 4, p. 234).

The experience of AGRARIA in Chile is an especially interesting case of this combination of on-farm and off-farm perspectives because it also shows how the NGO expanded its focus in response to changes in the state sector. Prior to 1990, AGRARIA had preoccupied itself with technology adaptation and extension, with a particular interest in wheat-based production systems. This was a role understood explicitly as meeting a need (technical assistance to the campesinos) unattended by the Pinochet regime. With the transition to democracy the new government has assumed responsibility for financing campesino technical assistance and will meet this by contracting other organizations (among which are NGOs) to perform the role (see Chs 6 and 7, below). However, the state still has no agency providing specific assistance in post-harvest technologies, marketing and other off-farm issues. Consequently, AGRARIA re-evaluated how to use its own financial resources, and is increasingly directing these to work in off-farm stages of the food system, and in particular to post-harvest activities. Chilean tax law is such that if a farmer can sell wheat directly to the parastatal wheat purchasing agency (COTRISA), rather than via a grain trader, she or he can recover a value-added tax, as well as gain a higher price. For campesinos acting alone, this is usually beyond their possibilities, because of the small amounts sold, the distances to the selling point and so forth. AGRARIA has intervened in one region to assess the feasibility of playing a bulking, sales and tax recovery role for the small farmers. The most strategic concern is to persuade the agency to extend its presence to additional areas where campesino production is significant (Aguirre and Namdar-Irani 1991, and personal communication, 1991). Initial experiences have been favourable, and suggest that the activity might both cover the costs AGRARIA incurs as intermediary while also increasing farmer incomes. In addition, the increased income to the farmer facilitates the adoption of the wheat technologies that AGRARIA has screened, validated, and which it promotes in its technical assistance work (see Annex 2, Profile 1, AGRARIA).

Extension methods

Many NGOs originated in Freirian popular education (and many others have flirted with it), so it is not surprising that one area in which they have introduced innovations has been in approaches to extension and rural education. These experiments have generally had one or several of the following aims: to make extension more participatory; to convey messages in local idiom and local practice; to target particular groups; to link extension practice to more general organizational strengthening; to render extension more efficient and effective.

Extension and rural education

Prior to Cotlear's and Figueroa and Bolliger's findings (Cotlear 1986; Figueroa and Boliger 1985), many NGOs were already seeking to link agricultural and educational work in rural areas. For the Colombian NGO, FUNDAEC, this was its *raison d'être* (see Annex 2, Profile 6). The focal point of its work has been the concept of the 'Rural University', an attempt to take high school and tertiary education out of urban areas into the countryside and find meeting points between it and traditional knowledge. The programme stems from an institutional conviction (paralleling the ideas of Schultz 1975) that formal education has a central role to play in rural development and is perhaps a cornerstone of technology transfer. During almost two decades, FUNDAEC developed a programme called the System for Tutored Learning (Box 5.5). This has twenty-eight texts developed by FUNDAEC which rural people (mainly young adults) work through in groups over periods of one and a half to two years. On the basis of its experience, FUNDAEC now works with texts covering five main areas (mathematics, science, technology, language and community service). The material covered is also linked to, and builds on, FUNDAEC's own ATD work. The supposition is that the students benefit directly as farm resource managers, and also go on to act as paratechnicians in their communities. Some of the graduates of these courses go on to higher-level courses which FUNDAEC organizes, and indeed some of these more advanced graduates have themselves subsequently formed their own, rurally based NGOs to provide direct technical assistance in their communities – an interesting instance of scaling-up (Correa, personal communication, 1992).

This combination of technology transfer and more general educational activities in rural areas has been noted in other NGOs (IFAD 1991; Carroll 1992). Indeed, in Colombia, FUNDAEC's experience has been so successful that the Ministry of Education and other NGOs have entered agreements with FUNDAEC for a 'scaling-up' of the project into other parts of Colombia. In this agreement, FUNDAEC advises these NGOs and provides teaching material. Such experiences are interesting because they show institutions combining the types of work (i.e. education and technical assistance) that Cotlear's material suggested must be combined in order to close the technology gap (Cotlear 1989a). This is an integration that has been lacking in the ATD work of NARS.

Forming campesino paratechnicians

The experience of FUNDAEC has linked general rural education with the training of farmer paratechnicians (CIED in Peru has a similar programme). The principles behind the training of paratechnicians are several:

- Farmer-to-farmer extension means that messages are transferred in local language, local idiom and indeed in everyday life.
- Paratechnicians are a cheaper form of extension agent.
- Paratechnicians are more likely to stay in communities than are extension agents, both in the short term (e.g. weekends) and in the longer term.
- Work with paratechnicians is part of a more general empowerment of rural people, demonstrating that they are able to do extension work as well as formally trained extensionists.

Among NGOs, one encounters different modes of working with paratechnicians. In some cases, they are trained to work as extension agents, with a small wage from the NGO. In other cases there is no wage, and the intention is that they will act voluntarily, charge for services or charge commission on products they sell (e.g. veterinary medicines). Usually they do not become staff, although in the case of CAAP in Ecuador, some indigenous people were hired, trained and employed as staff to work as community promoters.

In other cases, NGOs train farmers in certain techniques and the farmers make the commitment to return to their communities and install demonstration plots on their own farms. In Ecuador, for instance, in the area of Cumanda, FEPP trained farmers in a range of agroecological techniques. They then installed plots at home, and were willing to discuss them with anyone in their communities.

These strategies have been an approach to extension that NGOs use quite widely, through not at all universally. It is more frequently used by membership support organizations than by non-membership ones (Bebbington 1991). El Ceibo is one of the most sophisticated examples of a membership organization with a research and extension strategy staffed by campesinos. The level of training they have received surpasses that of most

Box 5.5 FUNDAEC and rural education

The System for Tutored Learning (SAT) is a formal, out-of-school based education system developed by FUNDAEC to satisfy the educational needs of young farmers and the exigencies of an alternative approach to development. Among other things, the SAT constitutes part of FUNDAEC's technology-transfer work.

The SAT offers three levels of qualification: Promoter, Practitioner and Batchelor's in Rural Welfare. The lowest level (Promoter) is the basis of the whole system, seeking to consolidate in young people the skills, abilities, knowledge and attitudes required for work in community service, agricultural and livestock technology, mathematics, sciences and languages.

Teaching/learning methods combine the development of work guides, linking FUNDAEC's theory and campesino practice. These are discussed in class and acted on at a field level. Curricula are also devised at a field level, and teaching is done in local communities by tutors trained by FUNDAEC. Most work links research and action.

In several parts of the course, students analyse and apply ideas FUNDAEC has generated in its own research and in projects dealing with campesino production. For instance, part of the Agricultural and Livestock Technology Unit deals with FUNDAEC's concepts of production sub-systems and elements of an alternative technology. The Unit also encourages students to develop and test their own concepts for use in the conditions of their own communities.

In the more advanced levels (Practitioner and Bachelor) questions of species diversity and ecosystem management are dealt with. The teaching then encourages students to use these ideas to develop and test species mixes and agroecosystems that differ from those suggested by FUNDAEC.

Source: Blandon and Prager 1991: 24–6, in FUNDAEC 1991.

paratechnicians: they have attended courses at NARS and international centres in other Latin American countries, and some have been sponsored by El Ceibo to do university degrees in Bolivia. None the less, the experience is indicative of how much can be done in a strategy based on paratechnicians. In El Ceibo's case, the impact on local production systems has been impressive (see Annex 2, Profile 4, El Ceibo)

Group-based extension

As part of their wider concern to build and strengthen local organizations and solidarity, NGOs typically conduct extension with groups of farmers. This also helps to multiply the impact of their technical assistance and is thus a more efficient way of conducting extension work.

The sorts of groups NGOs work with, however, are quite varied. In part this reflects variation in local circumstances, in part differences of strategy. Many choose to work with communities that exist as administrative units – this is frequently the case in the high Andes. Others work with cooperatives formed under agrarian-reform legislation: these too are administrative units, though with a productive rather than a sociocultural basis. The principle in each of these cases is that such organizations are ultimately the vehicle through which 'the grassroots' will exercise its political voice, and so must be strengthened. In the Andean case there is also a cultural rationale – namely, that the community is a part of Indian identity and ought therefore to be strengthened (CAAP 1991; Torres 1991).

Other NGOs, however, operate with a more differentiated concept of the rural population, and argue that technical assistance must be carefully targeted to particular needs. Some work at a community level and also target more specific support to groups, such as women, within the community. Others work only with target groups. In Sauzal in Chile, for instance, AGRARIA moved from locality-based groups to targeted groups. Initially groups of between twenty and twenty-five producers were organized on a geographical basis. Subsequently, more specific 'interest' groups were established in order to refine the technical support they received according to the producers' particular circumstances. However, the functioning of these groups is complicated by their geographical spread, leading to difficulties in bringing producers together for meetings or field-work days. Also, because these producers come from different localities, they do not necessarily know each other: this impedes group dynamics. These two factors meant that the AGRARIA field team has had to lend strong support to these groups for them to survive (Aguirre and Namdar-Irani 1992: 17).

Which approach is more valid is a point of debate, and the answer perhaps depends on the overall goals of the NGO as well as local circumstances. Not to differentiate within the community runs the risk that technical assistance will be more appropriate for some than others, and so cause differentiated impacts: while the hope may be that benefits will 'trickle down' within the community, this can hardly be guaranteed. Targeting assistance more specifically is an important recognition of the differentiation that exists in rural areas and is more likely to deliver the support most appropriate to the target group. However, by benefiting some and not others, it runs the risk of weakening the grassroots organization as a political and representative body. Perhaps the more general and important point, however, is that NGOs have experimented with different approaches and much could be learnt for a national programme of technology transfer from recording and systematically comparing these experiences.

Forms of participation

We commented earlier that NARS have operated with a concept of participation in ATD that is constrained by their resources and their mandate to work only with technological concerns. NGOs, potentially, have the autonomy to go beyond this type of participation. At least in their rhetoric, their perception of development as a process in which human capacities and goals are realized gives rise to a wider conception of participation.

Incorporating farmer knowledge

Farmer participatory research should mean that the orientation, adaptation and subsequent evaluation of the technology-generation process all build on farmers' ideas. Once again, NGOs seem relatively strong in achieving this end, although the extent to which they do so varies, as do their methods.

Most commonly, NGOs precede their work with diagnostic assessments in the region, evaluating current farmer practices. They then use these as a basis for the elaboration of technologies which combine farmer knowledge with elements of practices the NGO aims to introduce. These approaches are quite similar on paper to those of public institutions. The difference, however, is that the more local focus and accessibility of the NGO, along with their contacts with local organizations, tend to make the NGOs' awareness of farmer knowledge more sensitive. The propensity of NGOs to adapt continually their strategies suggests that this is so (see next section).

Some NGOs have, however, gone further and grounded their ATD work in indigenous knowledge. CAAP's experimental station research was an attempt to build a research programme that aimed to understand the rationales behind farmer practices, and then improve indigenous techniques in order to enhance their contribution to meeting such rationales. This is a qualitatively different approach to one which tries to incorporate external ideas into farmer practices.

CESA in Bolivia similarly based its agroecological approach to stabilizing colonist farms on the basis of indigenous knowledge. Local people conducted a soil survey in Yucumo (in the Department of Beni), and used their classifications as the basis of an agroecological zoning of the area. Initially CESA had contracted external specialists to do the survey, but found that farmers could make little sense of it except that, as one campesino commented, 'Now we know the lands we were given are useless and that we are hopeless farmers' (quoted in Kopp and Domingo 1991: 16). By helping translate between local and scientific vernacular, the indigenous classification also helped to show farmers the meaning and relevance of terms in the outside study.

These positive comments, however, require two caveats. The first is that the efforts to ground an ATD strategy in indigenous knowledge have yet to yield many results; in CAAP's case because of the complexity of the research programme and in CESA's because days are early. The second is that farmer knowledge has, as we noted, often incorporated modern technologies: some NGOs given to an agroecological perspective find it hard to incorporate this farmer knowledge into their strategy. As in NARS, there is a tendency in NGOs, at both ends of the technological spectrum, to incorporate only the sort of farmer knowledge one wants to hear.

Creating space for farmer control

As we noted, the concept of participation in many NGOs goes beyond the incorporation of indigenous knowledge into ATD processes and tries to create spaces in which farmer management of that process is strengthened – either through representation in NGO decision-making or advisory structures, or through the promotion of farmer organizations who then take over the reins of a project when the NGO pulls out. These are efforts that link the idea of participation to wider concepts of direct and representative forms of rural democracy.

In this regard, the fact that many NGOs have their origins in syndicalist, cooperativist, liberation theological or Freirian traditions is important, for all share the concern to strengthen popular organization. This has led to agricultural development strategies, project methodologies and actions that contribute to coordination between individual producers and, subsequently, between communities. In such a context, seed and input distribution systems, irrigation development and management, and the installation of on-farm trials on land that is collectively owned or collectively rented and worked, have become interesting areas of action. By creating spaces where joint action is necessary (Box 5.6), the hope is to foster the formation of campesino organization (CESA 1980, 1991; see Annex 2, Profile 2, CESA-Ecuador; Profile 3, CIPCA, Profile 5, CESA-Bolivia, Profile 6, FUNDAEC).

Box 5.6 Promoting farmer organizations for self-managed ATD

CESA-Ecuador aims to foster the emergence of federations linking the communities with which it works in a given region. Such federations take on several functions, one of which is to coordinate relations with the NGO. Another is to become increasingly active in the administration of research and extension activities. Federations begin to organize the multiplication and distribution of seed among member communities, and in some cases form, with the NGOs' support, their own farmer paratechnicians, who give technical advice (with mixed success) in communities. One federation has gone on to manage a barley-processing enterprise that buys communities' grain and finances some of its own costs with the profits. Finally, these federations become political actors, exerting pressure not only over CESA but also over other agencies. At times this pressure is exerted to a degree that is difficult for CESA to accept and relations between the NGO and the federation can become more difficult when the federation begins to want to work in ways, or with other sorts of NGOs (e.g. politically or religiously different), with which CESA disagrees.

Source: CESA 1991; interviews.

In this sense, participation is not only about garnering farmers' technological ideas. Instead the concern is that farmers, and subsequently their organizations, participate in the design, monitoring and adaptive management of the whole project. Ultimately the aim is that a self-sustaining organization be established and that it then continue to manage the project after the NGO begins to pull out (CESA 1980; Garcia *et al.* 1991). The intention is that from then on the organization should look to NGOs and other sources for specific shorter-term forms of assistance as and when required. Indeed, in some cases NGOs support such organizations with the hope that they will exert more effective pressure than can the NGO to make public agencies deliver services to campesinos (Box 5.7).

Box 5.7 SEPAS: Creating peasant organizations to be vehicles of political pressure

In the San Gil region of Santander in Colombia, tobacco-producing campesinos suffered the double constraint of water scarcity and overdependence on a single crop. SEPAS therefore stimulated base organizations to pressure public institutions to respond to the need to install irrigation infrastructure and identify alternative crops to tobacco. Cooperatives were created in which SEPAS was both a member and a source of technical support. The cooperatives succeeded in pressing first the municipal government and then the departmental authorities to install irrigation – and in the end, the municipal government itself became a member of the water cooperative.

This success was a result of several favourable factors: the strength of base organizations and their long history of representation before the state; the local recognition of a crisis situation; and a relatively open political context.

Source: Bebbington *et al.* 1992: 56.

This organizational strengthening ultimately makes the NGO's own work more difficult as the organization begins to be more assertive towards the NGO itself, and to demand more direct control over project decisions and finances. This can be an extremely productive tension, but it can lead to conflicts which ultimately lead the NGO to withdraw from a locality. As one example of this, CESA-Ecuador withdrew from a project in the parish of Columbe in the central highlands because it was unable to resolve tensions over the control of project resources that had emerged between it and the campesino organization with which it was working.

Despite these problems, one of the most important lessons from these different experiences is that participation should occur at several levels at the same time. On the one hand it is a daily participation in the form of a more egalitarian relationship between field worker and campesino; it is also a periodic participation through formal meetings between field staff and local farmer groups; and finally, and perhaps most importantly, it is institutional participation. By 'institutional participation' is meant permanent campesino representation in the coordinating committee for the particular project. This representation gives voice and vote to the organization, thus giving it certain power to monitor, direct, criticize and vote against those elements of the agricultural project with which it disagrees. By combining participation with accountability in this sense there is greater likelihood that local ideas and needs will be acted upon. It is also a step towards local democratization rarely found in state programmes.

Project flexibility and adaptability

We commented in the preceding discussion that the institutional structure of NGOs facilitates a more adaptive and flexible management of local development. The prominence of farming and food systems approaches in the study NGOs is one example of such flexibility. Higher levels of local and institutional autonomy also allow the NGOs to adapt to problems as they arise – both specifically agricultural problems as well as others which are non-agricultural but central to local livelihoods. It would be quite unfair to expect NARS to respond to earthquakes or cholera epidemics, because these are not within their mandate. But it is significant that our case-study NGOs have been able to do so. CAAP was instrumental in rural reconstruction work after the 1987 earthquake in Ecuador, and many Andean NGOs responded to the cholera outbreak of 1990/1.[6] This responsiveness strengthens relations of trust and mutual respect with communities, the benefits of which are carried over into more specifically ATD activities.

Such flexibility favours an institutional learning process within the NGO. The relative stability of NGO personnel and directors also favours this and is in sharp contrast to the chronic instability of staff in NARS. People accumulate experience, and if they stay in the NGO, so too does the NGO. This increases the likelihood of adapting and not making the same mistake twice. Of course, institutional learning is still not perfect. Excessive stability of staff in directorships (a not infrequent problem) can reduce communication within the NGO. Also, the at times obsessive concern to *act* means neither resources nor time are given to more systematic and written-up learning. One of the NGOs' most often stated reasons for preparing case studies for this research was that it gave them an opportunity for a more systematic learning process.

The case-study Profiles in Annex 2 give several instances of such adaptability, but two cases stand out, not only for their demonstration of flexibility but also of the remarkable willingness of the NGOs to admit and respond to their earlier mistakes (remarkable because most NGOs are far less willing to admit such failures). In its early work with settlers moving from the highlands to the tropical forests of Beni, CESA's social, organization and awareness-raising work was successful, but the financial and technical assistance it gave to help settlers consolidate their holdings failed. As yields declined after settlers replaced the forest with monocropped fields, CESA lacked the technical capacity to help them. They soon abandoned their plots and moved further into the forest. CESA followed them, and saw the same problems emerge all over again. This time CESA responded by trying to initiate an agroecological alternative to settlers' monocropping practices. This time, however, recognizing its own technical limitations, its strategy was to contact a range of other specialist institutions to make their expertise available to CESA and the colonists it worked with. This time the intervention of CESA holds out more possibility of having sustainable impacts in local production systems (see Annex 2, Profile 5, CESA-Bolivia).

In Sauzal, in the semi-arid interior of Chile, AGRARIA similarly learnt and adapted as it progressed. In the early years of its work, insufficiently detailed diagnoses of campesino production systems led to mistakes in AGRARIA's research and extension work. Indeed, it made many of the same mistakes frequently criticized in government programmes, such as working with high-cost, yield-maximizing, commodity-based technologies that were inappropriate to farmers' needs. It also lacked the capacity to do properly the research it endeavoured to undertake. However, its own monitoring work and dialogue with campesinos allowed it to identify such failings. In response to these errors, AGRARIA increasingly operated with a systems approach to small-farm development, incorporating both on- and off-farm concerns into its interventions, and combining technical support with other activities such as training, credit and organizational development. Interestingly, the more AGRARIA began to operate with a concept of farming *systems*, and to work towards the goal of introducing structural as well as functional changes into these systems, the more the organization felt the need to collaborate with other institutions for research inputs and other forms of support (see Annex 2, Profile 1, AGRARIA).

The two examples demonstrate an institutional capacity to advance, 'learning-by-doing'. This can have the beneficial effects of responsiveness that we have noted, but it should be noted that it can also have quite negative results. The immediacy of farmer needs at times leads NGOs to intervene with technologies that are as yet not validated. In some cases, the NGOs may manage to get by with such a fusion (or confusion) of research and extension activities. In other cases they are not so lucky. It is a risky approach to ATD in which the costs of those risks are borne by campesinos – as we will see later.

Material impacts

Our focus on methodological and institutional issues is partly a result of the recognition that these are where NGOs are strong. However, it also reflects, quite honestly, the relative paucity and weakness of data on the impact of NGOs in closing the yield gap and in alleviating rural poverty. This does *not* necessarily mean there is no impact, and other studies have begun to show some evidence of positive effects (Carroll 1992). What it does mean though is that NGOs are weak in monitoring and evaluating their own actions (see also

Table 5.3 Introduction of new varieties of wheat in AGRARIA – Sauzal project

Sectors	*% of producers who sow new varieties*	
	1985	*1988*
Huerta de Maule	0	47
San Isidro	21	44
Name	10	67
Purapel	0	33
Sauzal	17	67
Vado de la Patagua	0	92
Puico	0	67

Source: Aguirre and Namdar-Irani 1992.

Table 5.4 Evolution of nitrogen fertilizer use and yields in AGRARIA – Sauzal project

	1984	*1985*	*1986*	*1987*	*1988*	*1989*	*1990*
Yield	9.5	13.0	8.7	–	16.1	16.0	16.7
Units of nitrogen/ hectare	17.8	20.1	19.2	20.6	–	–	27.3

Source: Aguirre and Namdar-Irani 1992.

Riddell and Robinson 1993). Assessing Peruvian NGOs, Torres (1991) concluded that this weakness was a result of their living off donations. However, it may also reflect a preference to hide behind rhetoric. Several of the NGOs in the study were under increasing pressure from their donors to be more systematic in such monitoring. This seems a very appropriate criticism. However, donors may have to be willing to give financial support for such monitoring work if they wish to see it.

More monitoring would probably be in NGOs' favour. Indeed, in NGOs that have monitored their operations, impact may be apparent. Tables 5.3 and 5.4 demonstrate preliminary evidence of impacts of the work of one NGO, AGRARIA, in campesino production systems, and in one of its projects CESA-Ecuador shows impacts leading to a 400 per cent increase in milk and wool production in highland sheep production systems.

Of course, NGO impact should not only be measured in terms such as these, and much conceptual and methodological work remains to be done on assessing more social and qualitative impacts. None the less, ultimately the rural poor want higher incomes and more productive and secure food production systems (Bebbington *et al.* 1992). Even strong NGOs still have a way to go to install monitoring systems that demonstrate this impact.

Influencing and obtaining support from government

A final capacity of NGOs, in which they have become increasingly willing to develop a strength, has been to draw down resources and expertise from government, to support their own work and campesino livelihoods. Chapter 7 focuses specifically on the mechanisms through which this has been achieved, and so we delay discussion of this until later.

NGO weaknesses

At earlier stages in this research we were criticized for suggesting that NGOs are a panacea to the ills of public-sector rural development; this is not our intention at all. We have already begun suggesting some limitations of NGOs. In this section we use an (admittedly somewhat extreme) case study to illustrate these weaknesses. Many of the weaknesses stem from the same institutional structure and sociopolitical trajectory that have given rise to other positive qualities in NGOs. For instance, while their small size may encourage flexibility and a work ethic, it also means NGOs are constrained in being able to capitalize on, or operationalize fully, their perspectives on development. Similarly while their sociopolitical origins have fed a social commitment to their work, they have also given rise to NGOs staffed by urban intelligentsia and social scientists who lack technical skills and can have difficulties talking with the rural poor.

Systems approaches and appropriate technology

Despite the commitment to the idea of farming and food systems approaches, NGOs' size makes it difficult for them to generate the amount of data necessary to describe and understand a system (Torres 1991). It took AGRARIA six years, many mistakes and the assistance of a doctoral researcher to arrive at the system specification they now have in Sauzal.

Similarly, the concern to address many issues at once can lead to a generalism that in turn can lead to the shallowness that comes from doing too little of too many things. This problem is part of a larger one. To do systems research properly, two key resources are essential: high-quality technical-research support and time. NGOs have little of either. Being small and generalist, they lack technological specializations necessary to address particular problems in depth. Being donor-dependent, their project cycles are generally of three years' duration, even if their perspectives are far longer term: three years is far too short to be able to generate and validate technologies through a systems approach. Their social commitment to 'do something' can also lead to frustration with slow research. Together, this can lead NGOs to release technologies before they are validated and thus to conflate adaptive and applied research with transfer work[7] – a recipe for the release of imperfect, inappropriate technology.

Campesino participation and rural democratization

Earlier it was noted that once campesino organizations become relatively strong, tensions can emerge between them and the NGO. These tensions ultimately point to a limit on how far campesinos can participate in the NGO. One of the main points of conflict concerns the moment at which project management should be passed fully to the grassroots organization. NGOs usually wish to retain control of projects longer than the organization wishes. NGOs argue (and many campesino leaders acknowledge, off the record) that local organizations require high levels of professional training and of internal coherence before they can take on the responsibility of project management. Conversely, the campesino organization may argue that the NGO is receiving money in the name of the campesinos and then keeping it to itself. It is also worth noting that although NGOs speak of the need for professionalization and maturity in the grassroots organization, their own excessively 'social' rhetoric has often kept them from giving the organizations the training in business management, accounting, banking and so forth that are necessary for the organization to take efficient control of projects.

The rush to show impact in projects is also an obstacle to participation. Participation takes time and, when concerned to show products at the end of a three-year project, NGOs can be as susceptible as the state to riding somewhat roughshod over 'participation' in order to get technology 'out'.

Some of these problems are indicative of the fact that NGOs are generally urban, middle class and non-Indian, and this gives rise to tensions between them and poor, rural Indians and campesinos (see discussion in Ch 2, above). They also reflect the fact that NGOs, despite their rhetoric, wish to retain the authority and power that comes from control over resources. This, ultimately, is a social and structural phenomenon. NGOs' power, and income, comes from controlling financial resources. They could be assured of neither if campesino organizations controlled those funds.

Given these problems, NGOs can hardly be seen as reliable voices for representing the rural poor. In his review of member NGOs of CCTA, Torres (1991) also draws attention to the lack of democracy in some NGOs. He comments that 'the internal dynamic of these institutions is highly varied: some are democratic, others are highly authoritarian. In such a situation', he asks, 'how can some of them possibly claim they represent or can interpret the concerns of "the people"?'. A survey of the member NGOs of a Bolivian network, PROCADE, was similarly circumspect about how far these NGOs were genuinely participatory (W. Gonzalez 1991) – interestingly some of those NGOs were the same ones criticized by the Rivera-Cusicanqui (1990) study discussed earlier on.

NGOs, then, may have taken the question of participation well beyond the form it takes in state programmes. But they too have social, cultural and political reasons for not being fully democratic, and for not being able to represent campesino concerns properly (see Box 5.8). So while we would still argue that NGOs are a force promoting democracy in rural areas, and in ATD, they too are institutions that themselves need to be democratized.

Consolidating civil society and rural democratization

Another element in what Fox (1990a) has termed 'rural democratization' is the consolidation of rural institutions able to contribute to the representation of rural interests. If they are one such set of institutions, NGOs have had considerable difficulty in attaining this consolidation. Their financial dependence is obviously one key factor that can make NGOs ephemeral. If funds dry up, so does the organization. Also the tendency of donors to give money in small amounts leads to the proliferation of many small organizations. More insidiously, the competition for authority within an NGO can lead people to leave and create 'their own NGO'. The competition for funds and status can be an obstacle to communication among NGOs. The sum effect is that the NGO sector as a whole is weakened: less communication means poorer work all round as NGOs unknowingly repeat the mistakes of others. Competition and rivalry not only weaken yet further their claims to be authoritative representatives of rural interests: they also stand in the way of the stabilization and consolidation of an institutional warp and weft in rural areas.

The problem of coordination and communication among NGOs was frequently stressed in the research, and no more so than in Bolivia where the proliferation of NGOs has been most extreme (Box 5.8). However, each country has its 'fashionable' areas for NGO activity. They are fashionable partly because their particularly severe poverty attracts donor funds, and partly because their proximity to urban areas is attractive to NGO staff who wish to continue living in the city.[8]

In such areas of especially high NGO presence, problems of coordination are severe. In the worst cases, NGOs may: compete for campesino clients, creating their own patronage ties with rural people; or work in the same locality with different methods; try and set campesinos against other NGOs, and end up either confusing farmers, encouraging them to trade one NGO off against another; or all of these. The story in Box 5.8 speaks for itself.

Finally, democratization implies not only the consolidation of rural institutions, but also that they engage in some form of contact with the state. We opened Chapter 2 with a quotation about the bias of NARS towards rich farmers: if NGOs are to play a role in correcting this bias (rather than simply sniping at it) then they will need some sort of interaction with public institutions. Yet few have been willing to do this – understandably so, perhaps given their past experiences, but none the less to the detriment of their contribution to a fairer rural development.

Box 5.8 Conflicts and competition among NGOs in the eastern lowlands of Bolivia

The colonization zones of San Julian and Berlin are located between 100 and 200 kilometres north-east of the city Santa Cruz de la Sierra, eastern Bolivia, and cover an area of over 5,000 square kilometres. Colonization programmes commenced in the 1960s and have taken the form of semi-directed state-sponsored programmes, together with spontaneous settlement. Total population is now estimated at 28,264.

The range and number of agencies active within San Julian and Berlin is considerable, including some five public-sector institutions, twelve NGOs and ten producers' organizations, as well as private commercial interests. Given their different ideologies and objectives and sheer number, this has led to several situations of inter-agency competition and, at times, conflict. Such conflict can take different forms and have different effects. The following are some examples:

- *NGOs undermine the legitimacy of state institutions* Colonization commenced with state-sponsored settlement programmes, implemented by the National Colonization Institute (INC). In 1979 USAID financed further colonization activities also implemented by the INC. However, due to poor administration by the INC, subsequent USAID funding was diverted to the private voluntary sector, with the establishment of a new NGO, la Fundación Integral de Desarrollo (the Integral Foundation for Development, FIDES) responsible for the implementation of further phases of settlement programmes. While such action may have increased the effectiveness of colonization activities, such donor support to an NGO is clearly an example of an NGO substituting the state rather than collaboration between public and private sectors, and thus undermines state activities and legitimacy in the zone.
- *NGOs undermine grassroots organizations* NGOs often disregard the existence of grassroots organizations in the area. The NGO CARITAS-Ñuflo de Chávez comes into conflict with the local agricultural cooperatives of CCAVIP (Central de Cooperativas Agropecuarias de Villa Paraiso Ltda) and CCAB (Central de Cooperativas Agricolas de Berlin Ltda), preferring to establish its own project groups in rural communities where local cooperatives already exist.

  The North-American-backed Centro Agrario Sindical (CAS) is an affiliate of the American Institute for the Development of Free Labour (AIFLD), whose activities have been noted as primarily political with links to the CIA.[a] CAS has similarly undermined campesino organizations, by establishing rival agricultural cooperatives where rural cooperatives already function.

  This has led to cases such as that of the community '3 de Mayo' where there currently exists: a local agricultural cooperative affiliated to CCAVIP; a second agricultural cooperative established by CAS; the agricultural group formed by CARITAS-Ñuflo de Chávez; one women's group affiliated to the Federación Departamental de Clubes de Madres; and a second women's group established by CARITAS-Ñuflo de Chávez – all in a rural community of less than sixty families.
- *Poor coordination among NGOs with similar objectives* In the Brecha Casarabe, the NGOs, CIPCA and SACOA, while sharing similar left-wing

oriented philosophies, previously lacked coordination and operated with a perceived rivalry for 'recipient' communities. This resulted in a duplication of activities and a sub-optimal use of resources.

Another recent development, also in the Brecha Casarabe, is the formation of the Union of the Cooperatives of the Brecha Casarabe (CCBC) supported and sponsored by international volunteers. These cooperatives and their central office now exist within the same communities as an organization formed by CIPCA in the early 1980s.

- *NGOs give conflicting technical advice to campesinos* Conflicts also exist between IVS-USA and the International Heifer Project, the latter having implemented livestock projects with the Union of Small-Scale Livestock Producers in the Brecha Casarabe since the early 1980s. Confusion and conflicts are now occurring because veterinaries from the two organizations are offering conflicting technical advice.

  This duplication of activities and total lack of coordination results in a sub-optimal use of resources, has created confusion, rivalry and disunity within communities, and is consequently counter-productive to both rural empowerment and efficient agricultural resource management.

[a]La Asamblea Permanente de Derechos Humanos 'Tendencias Sindicales en el Mundo y en América Latina' Santa Cruz, Bolivia.

Source: Ayers 1992.

Donor dependence

A final, critical, weakness of NGOs has already been implied but must be pointed out more explicitly: their dependence on donor money, goals and time cycles. The short cycle funding of donors militates against careful research and screening of technologies, and promotes efforts to show impacts; the 'buzz-words' of donors mean NGOs have to respond to changing demands on them (poverty, environment, women, etc.), and the overall dependence on external funding means NGOs tend to overstate what they have achieved by way of participation, impact, efficiency and sustainability, and to resist evaluations which could otherwise generate useful lessons for improving the quality of NGO work. This is quite understandable, but bodes ill now that larger multilateral and bilateral donors are running to NGOs. If claims about NGOs' strengths have been overstated, then the rush to exploit these strengths will lead to disappointment and disillusion.

The limits of NGOs: greenhouses in the Bolivian Andes – NGOs and the promotion of inappropriate technology[9]

The weaknesses discussed in the preceding section are illustrated in a study from the Bolivian highlands, showing how they led to a situation where NGOs promoted a technology generated without much participation, that was largely inappropriate for campesino circumstances and ultimately disadopted in many cases. The following is based on B. Kohl (1992).

The Bolivian high Andes (*altiplano*) is a particularly hostile and risky environment for its indigenous inhabitants, who have worked as agriculturalists for over two thousand years. The zone is dry and cold, with recurrent droughts and frosts. The inhabitants of the *altiplano* suffer chronic malnutrition. In pre-colonial times they had access to the products of different ecological levels (the 'vertical archipelago') but the sociopolitical conditions allowing this access no longer exist.

In response to these nutritional needs, and in particular to a severe drought in 1982–3, Bolivian rural development NGOs introduced protected horticultural systems (PHS) allowing vegetable production in adverse environments. In the past decade some fifty Bolivian NGOs have introduced systems of intensive horticultural production, primarily in greenhouses built of adobe, with plastic covers. A survey of forty of these NGOs shows that their projects have generally failed to meet their goals. Within five years of being built, less than 15 per cent – in some cases less than 5 per cent – of the greenhouses remain in production. A few projects have over 90 per cent of installed units in production, but these have provided continuous technical and material support.

Large quantities of money and scarce human resources have therefore been devoted to PHS to little effect – how was such an unsuccessful technology selected and why was it promoted? To answer this, it is necessary to look more closely at the ways NGOs selected, developed and transferred these technologies.

Technology selection

The criteria used by an NGO in selecting a technology are institutional as well as technical, and include organizational and political objectives, institution building and securing a funding base to guarantee project and agency continuity. As one NGO director explained, 'I'm not sure what real or productive effect the greenhouses have had . . . but they have given us a foothold in eleven communities'. The scale and visibility of PHS projects also appeal to donors: they are small, finite projects, easy to implement, relatively inexpensive and highly visible. They are also easy to quantify, and so allow simple measurement of the project's 'impact'.

The rapid implementation encouraged by donors and by NGOs' project-based funding cycles has led to the premature introduction of unproven designs and management systems.

Few NGOs have conducted serious experimentation on PHS, and there is little information available from public-sector research on which they might draw. Indeed, PHS projects have generally not differentiated between experimentation, demonstration and implementation. A major reason for working on several stages of the technology development process simultaneously is that

NGO funders will not generally support research, even in an applied form. Therefore, NGOs often try to incorporate research into their projects with the result that a *de facto* trial-and-error process operates as the new technology is introduced. An early project implementer explains, 'There was always support to build greenhouses or a windpump: but not to develop or perfect them. So we shuffled the development costs between various projects and were always under pressure to complete the work.'

The lack of funding for research is compounded by lack of adequately trained personnel. Added to that is the pressure on NGOs dependent on external funding to demonstrate that funders have had value for money. Indeed, much of what passes for research within NGOs is information generated for funders' evaluations, and since the outcome may affect job security and even institutional survival, the pressure to manipulate results is considerable.

A further problem with the dissemination of NGO research findings is that it is hampered by the competition between Bolivian rural development NGOs for resources and recognition. This competitiveness is manifested in restrictions on access to data and to analytical tools, such as computer programs, for analysis of costs and output in PHS projects.

Overall, the lack of available, rigorously tested information about PHS has led to the development of a 'folklore' about their supposed advantages: generalizations based on partial data.

Technology development

Many PHS projects were designed and implemented without a full understanding of the complexities involved: projects which aimed to improve the nutritional status of participants often failed to assess nutritional deficiencies and the cultivated area necessary to have an impact. Another common project goal was income generation: for the most part it was unrealistic to expect PHS to achieve this in such a high-risk, impoverished environment.

PHS projects have been plagued by technical difficulties. These include problems related to design characteristics, which resulted in poor thermal qualities, high humidity and low durability of materials. The lack of information meant that common early mistakes, such as building a PHS before ensuring water supply, were repeated many times over.

Generally NGOs presented PHS as productive systems which could solve existing problems, not as experimental designs to be tested. The urban-based technicians were never accountable for the technical success of the project in the same sense as the participants were responsible for the system's construction and operation.

Campesinos were rarely involved in the problem definition processes which led to decisions to implement PHS. Even less often were they involved in the design of the productive systems or their adaptation to specific local needs.

Project participants often feel as if the projects 'belong' to the NGO rather than to the community.

A campesino union leader stated that, with the PHS projects, 'NGOs are trying to bring in foreign agricultural policies which have nothing to do with what the people want'. Another confirmed, 'These greenhouses don't help us to get ahead. Eating a couple more tomatoes doesn't pay school fees for our children nor does it buy them shoes'.

Central to the weak impact of PHS is that *altiplano* campesinos are not solely agriculturalists: their survival strategy in the face of high agricultural risks and low expectations encompasses diverse other productive activities. Project implementers have not always understood the opportunity costs associated with campesinos' participation in projects. The marginalization of PHS projects from the mainstream survival strategies of the Andean campesino thus limits their impact. So does the fact that PHS are not traditional agricultural systems. Their labour requirements are quite different: rather than being seasonal activities, PHS require daily attention. Their central logic also differs: rather than diversifying productive strategies to minimize risks, PHS depends on maximizing production through intensifying horticultural practices.

Another problem was that many PHS projects were organized around group production. There were two reasons for this. First communal production fitted the political agendas of the NGOs. Second, many project implementers did not understand the traditional division between communal labour (used for infrastructural work) and family-based labour (used for agricultural production) within *altiplano* communities.

In some of the more technically successful projects, production quotas and product distribution are set by the project implementers, and the participants appear to fulfil the role of peons to a new patron.

Technology transfer

The introduction of new productive systems ought to involve technology transfer on four levels: (1) from implementing agent to beneficiary; (2) between campesinos themselves; (3) between those developing the technology and the public domain; and (4) within the NGO itself. Of these, project implementers focus primarily on transferring information to beneficiaries, mainly through publications, workshops and short courses. This form of transfer has not been consistently successful, the two most common problems being campesinos' reluctance to thin plants sufficiently or to destroy diseased plants if they are producing fruit: both practices are counter-intuitive. Flows of information among project participants and to other campesinos are generally not given much attention by NGOs and are often hampered by NGOs' restrictive operating rules.

There are two channels of technology transfer from the NGO to the public domain: formal and informal. The formal channel – seminars and professional journals – is generally inaccessible to NGOs, while informal channels have failed to meet the need for information. Lack of data – on issues from organic pest and disease control to economic analysis – was cited by twenty-nine of the forty NGOs in the survey as one of their major problems. Even within the NGO itself, the institutionalization of personal experience is partial and imperfect, aggravated by conflicting agendas in different parts of the organization.

Lessons from the PHS experience

The early diffusion of information about PHS was largely due to a handful of charismatic individuals who were enthusiastic about the system's potential. Project directors' decisions to adopt the technology were based more on the personal enthusiasm of implementing staff rather than on the (very limited) research into PHS. Under funding pressure to produce rapid results, consultation with indigenous groups was limited, consideration of the social context inadequate and methods non-participatory. This, combined with a lack of conceptual clarity about the problems, and inadequate resources to implement efficient technical solutions, resulted in a wave of PHS projects which generally failed to meet their goals.

In Bolivia, the technology development function carried out elsewhere by public research institutions and universities has largely fallen to rural development NGOs. Their limitations as research and development agents are thus of particular importance within Bolivia, as well as illustrating issues of wider relevance. The limitations that NGOs suffer as highlighted by the PHS experience are:

- inadequate human and material resources to perform research;
- conflicts between scientific research and institutional agendas;
- restrictive institutional information policies and inter-institutional competitiveness;
- inadequate consideration of cultural factors in the selection and implementation of new technologies;
- a lack of expertise within NGOs on policy issues which affect technology selection on both macro and micro levels;
- vertical, non-participatory methods of selecting, developing and implementing new productive systems.

NGO strengths and weaknesses: a summary

This overview of NGO work in different aspects of ATD suggests that they have much to contribute to future strategies of technology development. The size of the NGO sector; their poverty focus; the propensity of many NGOs to concentrate on adapting elements of modern technology to campesino reality; the focus on the locality; their decentralized structure; the strong (probably excessive) concentration in areas of extreme poverty unattended by the state; and the concentration of interventions in the adaptive and transfer subsystems of ATS, all suggest the important role NGOs could play in closing the technology gap. Indeed, many see their role as such.

At the same time, it is the case that much of the work on agroecological and farming systems approaches has been done in NGOs. Consequently, the adaptation and infusion of low-input practices into a new technological agenda could build on NGOs' accumulated experience to great effect. On the other hand, we have also suggested that their capacity to conduct this research is greatly limited by constraints that are in part due to resources, in part to the short-term and competitive nature of NGO funding, and in part to the action-oriented culture of NGOs.

Two dimensions of NGOs' strengths stand out above all: their capacity as relatively flexible institutions to adapt and respond to local circumstances; and their important role as 'capacity builders', fostering local organization, education and skills formation.

However, many of these 'comparative advantages' of NGOs bring with them corollary shortcomings. The relative smallness of NGOs (Carroll *et al.* 1991) limits their capacity to do backstopping research, and to gain access to such support: distance from research centres, lack of time, lack of contacts, and so on, are all constraints. Their approach to participation is a genuine contribution to strengthening rural civil society and rural democracy – but they too as institutions could be much more transparent, participatory and accountable to campesinos. Their poverty focus is laudable but tends to attract many of them to the same 'fashionable' impoverished areas. This overkill also leads to problems of coordination between NGOs. Their localized focus and lack of time frustrates their engagement in debates over the policies that ultimately determine the impact of NGO interventions. Finally, although the innovativeness of many NGOs has been apparent, so has the fact that for every successful innovation there are many more failures and mistakes. Too often these experiences remain isolated, leaving other institutions to discover the same wheel, or more frequently, to discover the same broken wheel.

It is not surprising that NGOs have failings, and it is important to note them, for NGOs should not be expected to do the 'magic' that many are expecting of them (Clark 1991). These failings point to the need for access to sources of technical, professional and informational support. They also imply the need for greater incorporation of NGOs into integrating mechanisms to aid coordination, communication, wider spread of information on NGO experiences and, crucially, access to policy discussions. This last theme is important not only in order to put NGO experiences to wider effect through informing policy with them; it is also important in order for NGOs to follow through on their criticisms of official agricultural development strategies. If they are to criticize, then they ought to be expected to find ways of making their criticisms hit home. If they do not, then their role in democratizing agricultural development institutions will be greatly circumscribed.

CHAPTER SUMMARY

We have argued in this chapter that NARS face a resource crisis that, while more severe in some countries than others, is placing increasing constraints on their ability to do basic and applied research. This crisis means that NARS will no longer be able to act in all sub-systems of the ATS, or work on as many crops and animals as they have before. If they carry on as before they will be spreading resources ever more thinly and end up doing nothing effectively. NARS need to concentrate their work and resources more carefully – and as they do so they must seek new partners who can concentrate resources in other activities in the ATS.

Because of their limited mandate, NARS can only engage in a form of participation which, while it may help generate more appropriate technologies for campesinos, does not increase their capacity to make full use of that technology.

NGOs, if far from perfectly so, are able to work with a broader conception of participation which allows them to build capacity at the grassroots and can remove some of the local barriers to technology use. They also have a body of experience and knowledge with lower-input technologies that, if brought to bear on the work of NARS, could help reorient it to meet the challenges of a new technological agenda. NGOs' work is concentrated in the adaptive research and transfer sub-systems of the ATS, and indeed they have sometimes had negative impacts in this work because of poor access to more vigorous research support.

It would seem that the quality of NGOs' work in ATD as well as their ability to reorient formal ATD work in Latin America could be enhanced by more contact and linkage with the NARS and other public-sector institutions such as universities. They may then be potential partners for NARS.

In an earlier chapter, we suggested how broad changes in society were creating greater space for a collaboration between NGOs and government in general. In the next chapter we focus on specific changes in the ATS that also seem to be leading in this direction and may therefore facilitate this potential for partnership. In so doing they may be laying institutional bases for a strategy of ATD with more potential to close the technology gap and respond to the new technological agenda.

NOTES

1 Such work does occur in NARS. Andean crops programmes recover traditional varieties, and integrated pest management and soil and water programmes are to be

found. The point is that they are far less important, powerful and resourced than mainstream breeding programmes.

2 It should be noted that 1991 saw a return of PhD level staff to IBTA in anticipation of higher, competitive wages being paid as part of a new World Bank loan.

3 Responding to this realization, and with the World Bank's prompting, IBTA will now reduce its actions to just six programmes from 1992 onwards.

4 This NGO was CAAP in Ecuador.

5 David Lehmann has denominated these NGOs as the 'informal university': part of the generalized 'informalization' of Latin American economies (Lehmann 1990: 183–5).

6 CAAP's rapid response to the 1987 earthquake greatly impressed the then mayor of Cayambe, Diego Bonifaz. When, in 1988, Bonifaz became Sub-Secretary of Rural Development and began to design a new programme of rural development, PRONADER, this experience with CAAP led him to believe NGOs should play an important role in implementing the programme (Bonifaz, personal communication to A. J. Bebbington, 1990 and 1991).

7 It is significant that the NGO with the most controlled long-term crop research (CAAP) funded this out of its own resources, not donor-tied funds.

8 Urban proximity may be the more important factor in many cases. In Bolivia, of 153 agricultural development NGOs identified in an FAO study, forty-nine were in the province of La Paz, while only thirty-two operated in the provinces with the most extreme incidence of poverty; and only eleven of the forty-nine in La Paz operated in such areas (FAO 1990: 10, 13).

9 This sub-section is based on Kohl 1991.

6

INSTITUTIONAL CHANGES IN THE LATIN AMERICAN ATS

Creating room for manoeuvre in NGO–NARS collaboration?

The previous chapter drew attention to strengths, weaknesses and differences of perspective in the efforts of NARS and NGOs to foster agricultural development among resource poor farmers. In this chapter our goal is to identify recent initiatives taken by NGOs and NARS to respond to their institutional and operational weaknesses, and to take advantage of each other's strengths. In the case of NGOs we also consider their attempts to change the focus of the work of NARS and so scale-up their impact on ATD. Many of these initiatives until recently concentrated on fostering new relationships within the NARS or among NGOs. In recent years, however, they have involved attempts to forge new relationships between NGOs and NARS. This has been so both within sub-systems of the ATS and between these sub-systems. These changes have been in part an act of policy choice by the actors involved, and in part a response to changes in the context of the ATS.

Once again, to generalize about these processes is rash, but certain patterns seem apparent. Among NGOs, early efforts focused on challenging the perspectives of NARS on ATD. Over time, however, NGOs have paid increasing attention to their own limitations as development institutions. This has inspired the creation of networks and consortia among NGOs. These networks have had functional goals (e.g. to share technical resources among NGOs and address problems of poor coordination with other NGOs and public agencies), and political goals (to overcome the political weakness of NGOs when they act alone).[1] There has also been a growing interest in obtaining access to NARS support. It appears that some of these changes within the NGO sector may be opening new possibilities for their negotiations with governments and bilateral and multilateral donors, as well as for a relationship with NARS which will allow NGOs to influence government actions.

At the same time, NARS have responded to the criticism of their failure to meet campesino needs for technology. These responses initially revolved around efforts to incorporate systems and participatory perspectives into the

ATD process. However, these early initiatives have often been undermined as a result of public-sector resource scarcity. This has led NARS to look now for new partners. These changes within the NARS have also opened doors which may allow NGOs the opportunity to exercise increased influence over the approach of NARS to ATD, and over the allocation of public-sector resources. Yet at the same time, these changes have presented NGOs with a number of quandaries which can frustrate such relationships.

To discuss the trends in the NGO and public sectors in each of the five countries would be repetitive and laborious. Instead we will begin with a brief review of ways in which NGOs have sought to address the failings they see in the NARS and in their own ATD activities. We then focus the bulk of the chapter on changes that have been occurring in the public sectors of the region. The chapter suggests that these changes are fundamental in character, and that they point towards future relationships between NARS and NGOs that will differ greatly from those of the past. In the next chapter we discuss mechanisms for organizing such relationships.

NGO INITIATIVES TO CHANGE THE ATS

Exerting pressure on the NARS and government institutions

Most NGOs have been critical of the public sector's approach to agricultural development (Box 6.1). Many of them made no attempt to go beyond the rhetoric of criticism. Others, however, have tried to exert more direct influence over government perspectives. These are cases, then, where there have been forms of *interaction* between NGOs and NARS which have not necessarily implied formal, far less collaborative, linkage mechanisms. We have already made some general comments in this regard in Chapter 3, which we elaborate here on the basis of the case-study material.

Under repressive situations it was almost impossible for NGOs to exert pressure on NARS through direct contact. Instead they sought to influence the approaches of NARS through their more general actions in society – through their publications and their contacts with donors, political parties and popular organizations. Accepting that they would be unable to influence the NARS of the dictatorships, they understood their work as the elaboration of methodologies for rural development and ATD that could one day be scaled up under democratic regimes (see Box 6.2).

When the period of dictatorship came to an end, some NGOs were happy to continue pursuing only indirect means of exerting pressure on governments, but others initiated more direct contacts at local and national levels. Many of the cases of operational NGO–NARS relationships discussed in the following chapter are examples of such contacts. They were part of NGO strategies that had a dual purpose: (1) to enhance the efficiency of the ATS through improving the relationships between sub-systems and between institutions

Box 6.1 NGO criticisms of public-sector agricultural development

Typically, NGOs have criticized the agricultural development models of the NARS in Latin America on a number of grounds:

- that their focus on export and commodity crops is inequitable and inappropriate for most resource-poor farmers on marginal lands;
- that the high external input technologies they work with are agroecologically unsustainable;
- that their research methods are inappropriate because they are organized by commodities and not systems, and are determined by researchers and not farmers;
- that their main contact is with medium and large farmers, to the exclusion of resource poor producers;
- that their institutional organization is rigid, centralized and highly bureaucratic.

(see Ch. 7); and (2) to exert an influence over NARS researchers and decision-makers, in order to orient public-sector research and extension more towards campesino needs.

The difficulty for NGOs is that these direct contacts with government agencies reduce the efficiency of their own activities. They absorb NGO time and money costs, and expose NGOs to the bureaucratic tangles of public-sector operations, and at times to politically inspired attempts to influence NGO projects. The result is to increase NGO dependence on the generally slower and more interrupted time cycles and decision-making processes of government institutions (Torres 1991).

It is on account of difficulties such as these that many NGOs remain sceptical of such relationships with government. One strategy, however, that some NGOs have used to avoid these problems has been to try and exert influence at a personal level through giving training courses open to, or specially for, NARS staff. Favoured themes in such training to NARS have included participatory techniques, agroecology and systems perspectives on agricultural development (Box 6.2).[2] The advantage of this strategy is that the NGO avoids making any of its field activities dependent on government decisions, project cycles or funding. The disadvantage is that the NGO has no means of ensuring that themes from the courses are taken up in NARS work.

A consortium of agroecological NGOs, CLADES, has recently taken this approach a step further by establishing an agreement with the agronomy departments in ten different Latin American universities. Under this agreement, CLADES and its member NGOs will work with these universities to incorporate agroecological perspectives and low-input production practices into the curricula of agronomy courses. Rather than retrain NARS out of certain Green Revolution ways of thinking, CLADES instead will try and

Box 6.2 GIA: an NGO's influence over NARS and public institutions through training activities.

The department known as the Area of Campesino Development Strategies (ACDS) was founded within GIA in 1981. This department saw itself as providing:

- support for campesino survival strategies under the prevailing monetarist model of the dictatorship years;
- experience in the use of a methodology which would serve as a basis for defining, together with other research institutions, a national agricultural development policy for the campesino sector once democracy was restored in Chile.

During this period, GIA adapted the concepts of appropriate technology, production systems, farm-level research and of the micro-region to the conditions of the Chilean campesino. The intention was that these approaches should be applied on a wider scale than merely within GIA.

In one of their efforts to multiply the impact and social utility of this conceptual, methodological and practical work, GIA gave professional training courses. Although these did not constitute a direct influence on the policies of other organizations (much less those of the state), they did offer the possibility of gaining influence through the staff of those institutions. Under the dictatorship, this training was given only to other NGOs, with the aim of strengthening their institutional capacities. Since 1985, GIA has hosted an annual course which runs for ten months (five days a month), for twenty-five participants from different NGOs.

Since the return to democratic rule, however, GIA has begun to give training courses to professional and technical staff from public-sector institutions such as INDAP, INIA and CONAF (National Forestry Corporation) in such areas as diagnostic methodologies, campesino systems of production, and campesino organization. It has also given training in these areas to the technical teams of the different technology-transfer companies that implement technical assistance programmes on contract to the state.

Source: Sotomayor 1991.

influence technicians before they reach NARS and other institutions (Adriance 1992; Altieri and Yurjevic 1991).

Addressing the limitations of NGO approaches to ATD

With changes such as the return to democracy, and the degree of security it has offered, NGOs have shown a greater tendency for self-criticism. This self-criticism may also reflect the accumulation of experience among NGOs allowing them to identify the limitations of the development strategies they had initially pursued. Either way, there has been an increasing concern to address the problems that are characteristic of NGOs' work when they act in isolation: limited technological capacity, poor coordination with other institutions, weakness in negotiations with other institutions (especially government) and limited ability to scale up their experiences and innovations. As strategies to address these weaknesses have developed, they have often brought NGOs into closer relationships with public-sector organizations.

In recent workshops, NGOs have noted their need for access to programmes of peasant-centred agricultural technology development to generate the technologies they lack the capacity to develop (Bebbington *et al.* 1992; IICA 1987). Some forms of technological support (particularly in areas of agroecology and low-input agriculture) are located in other NGOs and international NGO networks, but participants at these workshops also recognized that the public sector frequently possesses much of the requisite expertise.

Aside from the political obstacles discussed earlier, NGO access to these resources has been hindered by the complexity of government bureaucratic structures and procedures. None the less, a significant number of NGOs have found a way through this, and their experiences are discussed in the following chapter. Those able to do so have tended to be larger NGOs who can afford the time that must be spent in seeking out information and resources from government offices. For smaller NGOs, contacts with the public sector are often prohibitively resource intensive.

Box 6.3 NGO network functions

Four main roles that NGO coordinating bodies perform are:

- to provide a forum for meetings and the exchange of ideas – this has been the most successful function;
- to provide services to members (e.g. provision of training in management, publications, networks/newsletters);
- to mediate state–NGO relations. The state likes umbrella groups because they can help the state: (1) make contact and liaise with NGOs, (2) co-ordinate programmes with NGOs and (3) monitor NGO actions;
- to channel finances to national NGOs – this is often the case where an external agency gives a block grant that the umbrella group subdivides and allocates among members.

Source: Stremlau 1987.

One NGO response to this problem of access to other sources of technical expertise has been to create networks and consortia of NGOs (Box 6.3) (Stremlau 1987).[3] These consortia have a number of origins and exist on different geographic scales. The case studies include examples across the whole range. There are local consortia (Box 6.4), provincial and national (see Box 6.5) and Latin American (see Box 6.6). As they have continued to develop, these consortia and networks have begun to influence the relationship among NGOs as well as between NGOs and the public sector (O'Brien 1991). Indeed, public agencies and large donors have come to welcome the existence of networks because they can help make discussions between government and NGOs more efficient (Stremlau 1987). It is easier for governments to talk to one network than to a large number of NGOs.

Box 6.4 A local informal network in Bolivia coordinates and strengthens institutional actions

In its work with farmers in a colonization zone in the tropical humid forest in Yucumo in the Department of Beni, CESA-Bolivia recognized increasingly:

- the need for an agroecological approach to resource management in Yucumo;
- that neither it nor other institutions working in the region knew sufficient about the area or about agroecology to meet farmers' needs.

These concerns were discussed among NGOs and farmers in the region and it was decided to create an informal coordinating network, the 'Yucumo Inter-institutional', which included seven NGOs, a technical college, the parish church and various farmers' organizations. The goals of the network were:

- to achieve some homogeneity among the different institutions as regards terminology and methodology;
- to coordinate research and training programmes.

Agriculture, livestock and health committees were created. Each was given the responsibility to train community paratechnicians.

In July 1990, the network organized a symposium on agroecology to which experts from outside agencies were invited. The symposium helped foster greater contact among institutions through reciprocal visits. It charged CESA with the task of executing a participatory study of agroecological systems in Yucumo.

This informal network thus fostered a degree of coordination among NGOs and helped exert a demand on external institutions for technical support.

Source: Kopp and Domingo 1991.

Some of these consortia have sociopolitical roots, but have subsequently moved into technical areas, and have begun to assume a number of tasks, including those of technical support, coordination, policy negotiation and assisting their members in gaining access to technological and information resources in other institutions (see Box 6.3). Others have a more clearly technical origin, particularly around the related themes of agroecology, indigenous technologies, sustainable-resource management and agroforestry. One presumption underlying these networks is that the NARS which have a Green Revolution focus cannot provide the agroecological support and advice needed by the networks' member NGOs. Instead these networks seek to facilitate the communication of knowledge among agroecological NGOs, and between them and universities in Latin America and elsewhere. Once again, however, these networks have also taken on a negotiating role and some have sought to encourage NARS to incorporate the perspectives of agroecology and low-input agriculture into their own programmes (see Box 6.6). It is as yet too early to tell whether they have had any influence.

A further interesting feature of those networks with a technical or regional focus (i.e. less politicized networks), is that in some cases they can bring NGOs, public-sector and producer organizations together in the same network. The Red Agroforestal (Agroforestry Network) in Ecuador for instance involves both NGO and public institutions concerned with forestry. While the discussions between participants are not always cordial, the Network has provided a forum for them to come together, learn of each others' activities and move towards coordination (Moreno, personal communication, 1991).

Among the five countries, the use of networks is perhaps most advanced in Bolivia.[4] This appears to be in part an effect of the political circumstances of the 1970s and early 1980s, when NGOs sought means to strengthen and protect themselves. It is also a response to the large number of NGOs in Bolivia, and the emergence of the more opportunistic organizations that have emerged in the 1980s (see Ch. 3). Some NGOs have used membership of networks closed to other NGOs as one of several means of distancing themselves from these opportunistic organizations (another means has been to retitle themselves as 'Private Institutions for Social Development' or IPDS – Instituciónes Privadas de Desarrollo Social).

Two of these national networks of Bolivian IPDS – UNITAS and AIPE[5] – which were in part formed with more political goals, have each undertaken large agricultural and resource-management programmes that seek to coordinate actions among their member NGOs. In these programmes, the network has coordinated the design of the programmes, acquired and administered project funds, facilitated exchange of information and experience among participating NGOs, handled relationships with donors and negotiated with government institutions. The networks have also served to search out and disseminate technical information from other sources and thus absorb the costs involved. In the case of UNITAS for instance (Box 6.5), the network searched unpublished material at IBTA dealing with high altitude pasture management, and published it as a literature review available to its own members as well as the public. In doing so it also provided a service for IBTA, which lacked the resources to undertake such a task.

Over time, however, some commentators have begun to suggest that national networks are not the most efficient way of organizing and coordinating operational relations among NGOs. Thus the past few years in Bolivia have seen the emergence of networks at a sub-national scale. Some NGO activities suggest that the national networks are most suited to policy level negotiations with government and donors, and should leave networks organized at the level of provinces or 'Departmentos' to coordinate local NGO actions and engage in more operational discussions with local branches of government. Such a provincial forum could allow member NGOs to collaborate in devising alternative, regional agricultural development policies

Box 6.5 NGO networks in Bolivia: coordination and negotiation in ATD

National NGO networks have existed for over a decade in Bolivia. In the 1980s two networks initiated agricultural programmes, one of which is the *Programa Campesino Alternativo de Desarrollo*, or PROCADE, of the network UNITAS. By 1991 it embraced fifteen NGOs working in 322 communities across five highland departments, drawing on a field staff of around forty-five agronomists and forty-five *educadores*. Building on an earlier programme (PRACA), PROCADE was launched to coordinate NGO research into agrarian alternatives. A plan for 1989–91 was devised, research themes divided among the member NGOs, and PROCADE channelled funds and coordinated internal publication and distribution of results. By coordinating research, PROCADE aims to add coherence to disparate NGO activity, set certain research and policy norms across a wide region and improve inter-NGO information exchange: a response to the problems of NGO proliferation.

PROCADE's research revolves around four programmes based conceptually in agroecology, farming systems, food security, and self-management. The goal is to adapt technology to regional circumstances; thus in Potosi and Oruro, research is conducted in improving indigenous subsistence technologies, while among more commercialized campesinos in Tarija, the focus is on modern inputs and mechanization.

The programme was a response to an NGO weakness and a NARS weakness. On the one hand, NGOs involved in relief work after the 1982/3 drought in highland Bolivia were faced with the need to go beyond relief to development actions. The activities of PRACA and PROCADE represent this move into development. On the other hand they are also a response to the weakness of IBTA in highland campesino communities. The goal, however, is not to replace IBTA, but rather to use PROCADE's strength to reorient IBTA so that its research is more appropriate to campesino needs, and to exploit PROCADE's coordinating role as an interface between IBTA and individual NGOs.

Sources: Gonzalez 1991; Miranda, personal communications, 1990 and 1991.

to present to the Bolivian state, as it turns to NGOs as implementers of agricultural programmes. By strengthening the networks' power of negotiation, this could promote decentralization of government agricultural policy-making and programmes, facilitating NGO influence over their design and implementation so that local campesino needs are more adequately addressed (Quiroz, personal communication, 1990; Bebbington 1991).

As coordinating mechanisms, networks and consortia are not, however, without their problems. They themselves generate bureaucratic costs and have been criticized for not being sufficiently democratic. The workshop discussing this research suggested that networks offer much potential, but their ability to realize it has yet to be proven. In some cases they can simply become a conduit for channelling donor funds to members, while excluding other NGOs from access to these funds by controlling membership of the consortium. Whilst these may be particular problems in the case of networks created at the instigation of governments or donors, NGO created networks are not exempt from such shortcomings either. Such networks can rapidly gain unprecedented power (funds, privileged access to government, national prominence) while not representing a large part of the NGO community. Consequently, donors should be circumspect about creating networks for their own purposes.

Notwithstanding these caveats, the origin of the network idea comes largely out of NGO experience, and its existence reflects an attempt to respond to particular NGO weaknesses. Assuming that networks' internal processes can indeed be made more efficient and democratic, then they may become an important institutional mechanism for the coordination of NGO–state–donor relationships (O'Brien 1991).

Box 6.6 CLADES: a Latin American NGO consortium

The Latin American Consortium for Agroecology and Development (CLADES) is made up of eleven NGOs from eight South American countries (in 1991). Towards the end of the 1980s, many agroecological NGOs were seeking ways to achieve a greater impact in terms of sustainable rural development; CLADES was formed in 1988 with this object in mind. CLADES proposed that, in order to achieve this aim, the following would be of major importance: (1) an improvement in collaboration between NGOs, governmental organizations, academic and international institutions; (2) a strengthening of the agroecological perspective in these institutions; (3) a change in relations between NGOs and other institutions in Latin America and the international development agencies; (4) a strengthening of NGOs' institutional, technical and professional capacities; and (5) a division of labour between the different institutions according to their comparative advantages.

On beginning this undertaking, one of the main activities was to synthesize and diffuse information on agroecological innovations and work that was held in the many different and often isolated institutions in Latin America and (especially) the United States. Subsequently, this information was used for professional training activities organized and coordinated by CLADES. This training, although in large part aimed at NGOs, was also open to technicians from the state sector. In organizing these training activities CLADES' aims have been to foster coordination and communication between different types of institutions, and to strengthen the professional capacity of NGOs, above all in their use of an agroecological perspective.

At the same time, CLADES has organized seminars for NGO executives to discuss amongst themselves how best to respond at an institutional level to the politico-economic changes that are pressing NGOs to take a more prominent role in development.

Source: Altieri and Yurjevic 1991.

NGO initiatives: a summary

Overall then, the trend among NGOs seems to suggest a shift away from criticizing government at a distance, towards an increasing awareness of the limitations of the NGOs' own approach to development (particularly when the NGO works alone) and a growing concern to upgrade the technical expertise of the NGO. This has led to a number of initiatives that reflect an increased interest among NGOs to end the distance between themselves and government, and to begin investigating the feasibility of contacts with public institutions. Some of these initiatives have led individual NGOs to open relationships with government institutions, others have involved greater contact among NGOs in coordinating bodies that have themselves initiated contacts with governments. In all cases, there are still many imperfections in how these relationships function. However, they are significant changes not least because they parallel a number of similar trends in NARS which reflect a growing self-criticism in the public sector and an increased concern to initiate contacts with NGOs as one means of responding to recognized weaknesses.

NARS INITIATIVES TO CHANGE THE STRUCTURE OF THE ATS

In very broad terms, we can divide into two categories the different responses of NARS to the criticism that their work has excluded the rural poor, generated inappropriate technologies for them and been largely inefficient and ineffective in rural poverty alleviation and campesino development. On the one hand there are those responses that occurred in conditions of relative adequacy of public resources. These changes sought to respond through *internal* reorganizations of research and extension. On the other hand there are those that have occurred more recently in policy contexts in which governments claim that there is scarcity of public-sector resources. In these cases, responses have been sought through a combination of internal changes and a reorganization of the NARS relations with other actors in the ATS.

Responses under conditions of public-sector resource availability: systems and participatory perspectives within the NARS

Where resources have been relatively abundant, the typical response of many developing countries' NARS to criticisms of their Green Revolution focus been to incorporate systems, and subsequently participatory, research perspectives into the ways they conceptualized and organized the process of ATD. They have also experimented with institutional reorganizations in the public sector, successively separating and uniting research and extension (Kaimowitz 1990). These reorganizations have sought a balance between increasing the level of feedback from extension into the research process and avoiding excessive bureaucratization and institutional complexity (Merrill-Sands and Kaimowitz 1990; Kaimowitz 1990). These changes aim to reorganize the relations between public-sector actors in different sub-systems of the ATS – that is, between basic researchers, applied and adaptive researchers and transfer agents.

These changes have been strongly influenced by contextual factors. Two of the most important of these were donor support for systems and participatory approaches to ATD, and a political context that at the very least was not strongly opposed to the pro-campesino option implied by these perspectives. At times, political contexts were such that no amount of donor interest would lead to these changes. Certain regimes in Bolivia and Chile, for instance, had little interest in campesinos (Carroll 1992), and there was no real political support for the creation of such programmes. Consequently, in IBTA, although certain programmes have been oriented to on-farm and systems perspectives for some time, it was only in 1990 that the institution as a whole began a more concerted attempt to incorporate systems methodologies into its research activities (Zambrano, personal communications, 1990). Similarly, only in 1990 did INIA in Chile adopt a small-farm systems and on-farm approach to ATD. We return to these changes in the following section.

Elsewhere and at other times when there was some political support for the idea of a more participatory and systems oriented ATD, this has been attempted in several ways. In certain cases, as in the case of the PIPs in INIAP, a separate programme was created within the NARS (Box 6.7). In other cases, particular commodity programmes sought to adopt a more systemic on-farm approach. This has been particularly the case in the Andean crops programmes in each of Ecuador, Peru and Bolivia, reflecting the fact that these largely subsistence, undervalued crops grown by indigenous resource-poor farmers tended to attract researchers with a greater social commitment to the Andean campesino.[6]

In certain cases, the existence of links between NARS and Integrated Rural Development Programmes (IRDPs) with participatory and integrative perspectives on campesino agriculture also helped promote a more participatory mode of ATD. This was, for instance, the case in Colombia (Estrada 1991; Engel 1990) and in Ecuador (Cardoso 1991; Soliz *et al.* 1989). These projects tended to work with farmer groups at an on-farm level. They were also better able to link actions in the adaptive research and transfer sub-systems by combining interventions in these two sub-systems within the one, well-resourced project. In some cases such projects have achieved significant yield impacts among small farmers (Engel 1990).

There were, however, obstacles to the sustainability of each of these institutional arrangements. Some of these obstacles originated in the contexts

Box 6.7 The PIPs in Ecuador: a NARS attempts to institutionalize an on-farm research programme

The Production Research Programmes (PIPs) of the National Institute of Agricultural Research (INIAP) were set up in 1979, following positive on-farm research experiences in collaboration with CIMMYT. The PIPs constituted the adaptive research stage in INIAP's work. In them the technology developed in research stations was adapted to the agro-socioeconomic conditions of small and medium producers in a specific location. Each PIP existed within a province, employing one or two agronomists. They worked with a partial systems approach, that is one that identified production systems fairly rapidly and tried to improve two or three principal components. Neither sufficient personnel nor resources exist for a purer systems approach.

Campesino participation in PIPs research has normally been restricted to the preparation, installation and harvesting of trial plots. Participation in their planning or evaluating trials has been infrequent. None the less, in some cases working with campesino organizations has led to increased participation at the instigation of the organizations (e.g. in the Chimborazo province).

As well as undertaking adaptive research and testing technologies, the PIPs aimed to serve as bridging mechanisms in several ways:

- between field conditions and the researcher in the experimental station, so that the latter could reorient the research on the basis of feedback from the PIP;
- between some campesinos and the technology-generating process;
- between the generation of technology and transfer agents at local level.

The PIPs have suffered from various problems. These have limited their impact and their success in fulfilling a bridging role.

Their role in reorienting the work of the experimental stations has been blocked because PIP technicians have lacked the necessary influence, and also because of the intransigence of the more orthodox researchers within these stations. This relationship has gradually improved due to the increased presence of technicians in the experimental station and to the extent that joint meetings are now held.

As far as regards their role as coordinators of public-sector technology-transfer agents, experience to date has demonstrated the need to strengthen the capacity for coordination and regulated interaction in any future work. The PIPs' main counterpart has been the Ministry of Agriculture and Livestock (MAG). The bureaucratic procedures of MAG (such as fixed working hours), its permanent ongoing restructuring, and methodological differences have created problems in relations between the PIP and the extension sector. Similarly, when three PIPs were linked to rural development projects – DRIs – (to help them in technology development), inter-institutional competition, frequent and over-politicized changes of DRI directors, combined with the institutional rigidity of the DRIs did not allow the flexibility and continuity necessary to carry out the research, and in two out of the three cases the PIPs pulled out of the DRIs. These experiences show the importance of having agreements which permit, and indeed which demand, increased coordination between on-farm research and technology transfer.

Source: Cardoso 1991.

of the programmes. The creation of separate programmes of on-farm, systems-based research was resource-intensive and generally dependent on unsustainable levels of donor support. Similarly, when there were successes in IRDPs, these were made possible by high levels of donor support that were for fixed periods: once project funding came to an end, so did the successful reorganization of the relationship between actors and sub-systems. Furthermore, in the case of the PIPs in Ecuador, the political support at the creation of the programme declined under the rightist government of 1984–8, resulting in loss of resources and an extremely difficult period during which the PIPs became largely paralysed (Soliz *et al.* 1989; Biggs 1989).

Other obstacles lay within the public institutions themselves, stemming from problems of competition and relationships of power. Established station researchers frequently took little note of the usually younger professionals in the new systems and on-farm research programmes, thus frustrating 'feedback' effects. Also, some of the new inter-institutional and inter-programme relations created jealousies and competition over resources within the public sector (Box 6.8) (see Merrill-Sands and Kaimowitz 1989).

Changes in the NARS in contexts of resource cutbacks: opening up to other actors

The difficulty of institionalizing systems and participatory perspectives in NARS has further been aggravated by the pressure to reduce expenditure in the public sector. This makes it increasingly difficult to sustain even orthodox ATD work in NARS. As NARS and donors have looked around for ways of sustaining ATD with reduced public expenditure they have shown much interest in the increasingly important and large NGO sectors in these different countries. As a result, over the past few years proposals have come thick and fast for closer collaboration between NARS and NGOs.

Proposals are presented as if they were primarily motivated by the concern to orient ATD more towards peasant farmers. That they have this concern is undoubtedly true, but the reticence to acknowledge the parallel goal of reducing government expenditure has caused considerable annoyance among NGOs, engendering an often cool initial response to these proposals. Moreover, the participation of NGOs in the formulation of these proposals has generally been limited, which has been another source of conflict. The only country that is somewhat different in this regard is Chile, where, although concerns to control public expenditure are one factor motivating proposals for NGO–NARS collaboration, more important has been the genuine concern to build lessons from NGO experiences into the reorganization of the NARS, and to reorient public-sector contribution to ATD and rural development to address far more forcefully the problems of the rural poor.

None the less, and notwithstanding the difficult context of structural adjustment, the changes underway in the region have the potential to enhance the democratization of public-sector research and technology transfer. Indeed, in several cases NGOs have been able to take advantage of these changes and use them to press for an increased role in resource allocation and decision-making in any reconstituted ATS. The following sections discuss these changes in several countries.

Chile

In the period since the return to democracy, there have been profound changes in INIA and INDAP (respectively the research and extension institutions of the Chilean NARS). Of all the countries studied, it is in Chile that there has been the most rapid increase in the participation of NGOs in both implementation and decision-making in the public sector's ATD activities, though this has not been without its problems (Aguirre and Namdar-Irani 1992).

This high degree of NGO involvement reflects the fact that these sectoral changes have themselves been an integral part of the process of post-Pinochet reconstruction, and that professionals from the NGOs have played important roles in this whole process of democratization and institutional change (Box 6.8). NGO staff worked in the agricultural commissions of the two main opposition parties (the Party for Democracy and Aylwin's Christian Democrats), and since the election several of these professionals have moved from NGOs into government positions (Berdegué 1990; Sotomayor 1991). NGOs and their individual staff members have also continued to participate in the design of agricultural development programmes. Throughout, they have brought their prior experiences in NGOs to the recommendations they make for institutional, programme and policy change (Boxes 6.8 and 6.9).

Box 6.8 NGOs and policy-making in Chile

'The program [of the Aylwin government] was devised by party leaders, intellectuals and professionals who had opposed the military regime. Almost all of them were also affiliated with a network of NGOs that had expanded dramatically under the military dictatorship' (Loveman 1991: 8).

One factor that has aided the comparatively close relationship that has emerged between INIA, INDAP (respectively the public-sector research and extension institutions) and the NGOs has been the personal contacts among professionals in different institutions that date from their work in NGOs and political parties during the pre-democracy period. Another factor is the respect that NGOs have for the institutional and technical capacity of INIA and INDAP – a situation that stands in contrast to that in other countries where NGOs continue to criticize the weakness of their public sectors. While other economies in the region stagnated in the 1980s, the Chilean economy under Pinochet continued to grow. Consequently public institutions did not decline as severely as in the Andean countries, and INIA and INDAP consolidated significant professional expertise and institutional resources. Now, since 1990, INIA, INDAP and the public sector in general have been able to capture more bilateral and multilateral resources that have been used to support, among other things, rural development and welfare programmes.

A further crucial factor in the relatively good NGO–state relationship in Chile has been a certain convergence of opinion on the role and responsibility of the state towards the rural poor. In crude terms, the state has acknowledged an inherited social debt to the poor, and has accepted that it has a role to play in poverty-alleviation programmes. At the same time, many NGOs accept that the state is not necessarily best equipped to implement programmes of social and agricultural development, and that the implementation of these tasks are at times better suited to private agencies such as NGOs.

Professionals from several NGOs worked as consultants on the redesign of INIA, an institute formerly oriented mainly towards large, commercial farmers (Barril and Berdegué 1988). The goal has been to reorganize it so that in the future it also takes on the responsibility of generating technology suitable for small farmers. The main element of this change has been a new regionalized, on-farm research programme based on so-called Centres for the Adjustment and Transfer of Technology (CATTs) in which experiment station technologies will be adjusted to small farmer conditions (Sotomayor 1991). NGOs have been involved in the location of these. More importantly the concept of the CATT allows for the participation of representatives of INIA, INDAP, farmers, NGOs and private extension consultancies in committees that will monitor the research programmes and functioning of the CATTs (Box 6.9).

The move to increase scope for NGO involvement in INIA was paralleled by efforts to do the same in the technology-transfer programme in INDAP. This shift has involved fewer structural changes than the reorganization of INIA, as government had already withdrawn from direct provision of technical assistance to farmers, and INDAP's approach has been for some time to contract private-sector 'technology-transfer companies' to provide this service (Wilson 1991). Under Pinochet, NGOs were not considered eligible for such contracts, but with the new government they have been allowed to apply and have become one of the principal recipients of these contracts. In 1990, 45 per cent of the 42,000 small farmers receiving INDAP assistance were serviced by NGOs, and 12 per cent by farmer organizations (Ortega, quoted in Sotomayor 1991: 22). The rapid rise of NGO involvement, however, reflects the fact that NGOs have simply been incorporated into the same purely implementational roles as the private 'technology-transfer companies' did in the past. In this

Box 6.9 Opening INIA to NGO participation in Chile

The post-democratization changes in Chile have led to several fundamental changes in INIA, involving:

- a shift towards on-farm experimentation, with a corresponding reduction in on-station research;
- a change in research priorities: an important portion of the budget is now allocated to the small-farm sector;
- an effort to identify research priorities by agroecological area, involving a prior requirement for technical and socioeconomic characterization of these areas.

Central to the change in INIA have been the provisions for creating fifty-four Centres for Adaptation and Transfer of Technology (CATTs). The overall goal is to conduct on-farm agronomic research in each of the agroecological areas for which each INIA experimental station has responsibility, and on the basis of this research to specify, evaluate and propose improvements for typical small-farm systems in each area. To respond to these challenges, INIA has signed agreements with INDAP, small farmers' organizations and NGOs. More specifically, each CATT has a committee that monitors the functioning and research of the CATT, and on this committee INIA, INDAP, farmers and NGOs (or private extension companies) will be represented.

Source: Sotomayor 1991.

sense, the early changes in INDAP did not open spaces for the participation of NGOs in the design of the INDAP programmes.

Thus, at the same time as NGOs participated in the INDAP programme, they sought to have more influence over its design. They criticized the way in which INDAP separated technology transfer from other aspects of food-system development (such as agro-industry, local organization, and marketing). They also criticized its extension methodology. For while NGOs such as AGRARIA and GIA typically work with farmer groups, INDAP initially insisted they should work with individual farmers.

By the second and third years INDAP showed increasing willingness to respond to these criticisms. It began to allow NGOs on INDAP contracts to conduct some of their work in group form. In 1992–3 INDAP will introduce a new programme for working with private organizations such as NGOs and farmers' organizations under which INDAP will co-finance programmes with NGOs and other groups. At least in its design the programme seems to be a response to NGO concerns that technology transfer have a food systems perspective (Box 6.10). Indeed, in return for a contribution to project costs, INDAP appears to be offering groups such as NGOs a greater say in the design of agricultural development programmes.

Thus, whilst there have been tensions in the relationships between the NGOs and the NARS, the Chilean case stands out as one in which (1) the sense of shared goals in rural development is stronger than any sense of antagonism between state and NGOs, and (2) NGOs have been able to influence the ways in which public-sector ATD programmes have developed.

Box 6.10 Co-financing in INDAP: the state incorporates NGO lessons

In the new programme for cofinancing projects between INDAP and private entities, the private entity (that which designs and implements the programme) will have to pay at least 20 per cent of the costs of the projects (INDAP 1992).

In part this might reflect an effort on the part of INDAP to gain access to, and exercise some control over NGO funds. However, it also appears to reflect an INDAP response to NGO criticisms, and a concern on the part of INDAP to learn from NGOs. Thus, while other INDAP programmes only involve technology transfer, in the co-financing programme INDAP will contribute to projects that combine transfer with agricultural, livestock and forestry development activities. In particular, INDAP claims an interest in linking transfer work to credit systems, campesino irrigation, agro-industry development, organizational strengthening and support to women and rural youth. It also states that in its selection of projects it will prioritize innovative projects that have the potential to generate useful lessons for subsequent INDAP programmes (INDAP 1992).

These innovations are significant. They suggest that, in accord with suggestions from NGOs and others, INDAP is going to adopt a food-systems perspective and consciously incorporate lessons from NGOs work. The design of this project, at least, bears the stamp of an NGO influence.

Bolivia

The changes in the Chilean NARS were an effect of democratization within a context where NGOs and state alike recognized the constraints on expanding public-sector programmes. In Bolivia, the changes have been much more directly due to public-sector resource constraints and less due to political advances in the public sector. Consequently, NGO response to the proposals, and NARS response to NGO criticisms, have been less positive than in Chile.

The nature of the changes, and the NGO–NARS relations that they have engendered, differed between that part of the NARS operating mainly in the highlands, IBTA, and that operating in the Bolivian lowland department of Santa Cruz, which is served by a separate government research institute, the Centre for Tropical Agricultural Research (CIAT). We consider each in turn.

IBTA in the Bolivian highlands

With the growing fiscal crisis of the Bolivian state, in 1985 the government introduced the so-called New Economic Policy which initiated a radical programme of stabilization and adjustment including cutbacks in government expenditure (COTESU–MACA–ILDIS 1990). In this context, and in response to an already very weak IBTA, an ISNAR mission recommended that the extension service in IBTA be closed, and that resources be concentrated on research activities (ISNAR 1989b). ISNAR argued that extension should be left to the increasing number of NGOs (already employing some 1,200 transfer agents) and other so-called 'intermediate users' – institutions using IBTA technology as the basis of the technical assistance they give to the rural poor (Bojanic 1991). These principles were endorsed in a World Bank report for a loan to revive IBTA, and subsequently accepted by IBTA. However, the proposed programme has had a number of weaknesses that have hampered links between NGOs and IBTA.

The main weakness has been that these plans were made without much initial consultation with NGOs, and in a context where informal contacts between IBTA and NGOs were weak, and mutual distrust lingered (Box 6.11). Most institutional decision-makers in IBTA had very little personal knowledge of or contact with NGOs. As a consequence, the attitude to NGOs in the early formulations of the proposal was a highly instrumentalist one – they were to be the vehicles for the transfer of IBTA's technology, but were not accorded positions to influence IBTA's agenda.

Not surprisingly, early NGO response to the proposals was sceptical, and often critical. Moreover, many expressed the view that IBTA was so weak that it was unable to offer any sort of support to them anyway. (IBTA was far weaker than INIA in Chile.) It was, therefore, not clear that the proposals offered NGOs anything at all: neither money, support, nor potential influence over public-sector decisions.

Box 6.11 Lingering distrust between Bolivian NGOs and the NARS

In Bolivia, the distance between the state and NGOs is greater in the Andean highland region than in Santa Cruz, given that to a large extent NGOs have developed in response to the weakness of the state or in some cases as opposition bodies. Here NGOs have become one of the few sources of employment for those who do not toe the line with the policies of the various governments, particularly in the case of the *de facto* regimes. This in turn has resulted in a reciprocal mistrust which has yet to be fully overcome, and which complicates relations between state institutions and NGOs. In Santa Cruz, where political repression was less acute, relations have been much easier to establish.

Sources: Bojanic 1991; Veléz and Thiele 1991.

As they voiced their criticisms of the proposals, several factors strengthened the negotiating power of a set of the strongest NGOs. Perhaps the most important factor was the overall strength of the NGO sector, due not only to its absolute size and resources, but also its *relative* strength *vis-à-vis* IBTA. The proposed changes in IBTA sprang from a situation in which IBTA was extremely weak and NGOs were strong. The planned reorganization was in many ways a formalization of IBTA's dependence on NGOs for the transfer of IBTA technology. Conversely, as NGOs had not previously used IBTA to any great extent they were in no sense dependent on IBTA technology and could easily carry on as before. This context reduced IBTA's capacity to dictate to NGOs, and meant that NGOs could try and use the moment to push for greater influence over IBTA (see Box 6.12).

At the same time, NGOs were both aided and persuaded in taking a critical stance by certain donors who believed NGOs could play an important role in improving the public sector but that the early proposals for restructuring IBTA did not take advantage of this potential. COTESU, the Swiss bilateral aid agency, for instance, supported informal discussion groups involving NGOs, IBTA and other donor agencies. These groups gave NGO representatives an informal forum in which to voice their criticisms of the initial proposals and of wider aspects of agricultural and rural development policy. COTESU then went on to support a consultancy to develop the theme of the relationship between research and technology transfer in the reconstructed IBTA, within which the issue of the role for NGOs was also given attention (Werter 1991). Once again, discussions around the consultancy report provided fora in which the issues could be addressed and elaborated.

Finally, the prior formation of coordinating bodies among NGOs, especially the creation of PROCADE, provided a relatively strong nexus through which NGOs were able to coordinate negotiations with IBTA.

It would be wrong to say that the negotiations were either easy or that they have allowed NGOs to achieve the degree of influence they desire. Furthermore, the changes in IBTA are still at an early stage. None the less, despite the criticisms that have been made of IBTA's proposals, NGOs have been able to develop operational relations with IBTA and gain some role in setting research and institutional policy (Box 6.12). This has been due to the combined effect of the institutional changes underway and informal contacts and relationships among staff of NGOs, IBTA and donors.

CIAT in the Bolivian lowlands

The recommendations made by ISNAR and the World Bank for the reorganization of IBTA were based on the ways in which research and extension had already been reorganized in the Bolivian lowland department of Santa Cruz. The context in which the changes emerged in CIAT were both similar (public-sector fiscal crisis), but also significantly different as regards the relationships among the actors involved (Box 6.13). These differences have led to relationships between NGOs and CIAT in Santa Cruz that have been far more productive than relations between IBTA and NGOs in the highlands.

In the model of ATD that emerged at CIAT in the late 1980s, a series of

Box 6.12 NGOs take advantage of a restructured IBTA to gain influence over NARS decision-making

Between 1975 and 1989 IBTA had a staff of 120 extension agents distributed in eighty offices in the highlands of Bolivia. Increasingly, however, the fiscal crisis of the Bolivian public sector led to declining resources to support even the most basic costs for the field work of these agents. The response has been to embark on a radical restructuring of the functional and institutional relationships linking research and extension. The extension service in IBTA has been closed, leaving IBTA to concentrate on research activities in six main commodity areas, with extension being left to NGOs who, it is proposed by IBTA, will receive IBTA's technical assistance and information. Proposals, however, have paid more attention to research than to extension and have failed to formulate clear mechanisms for linking the sub-systems of research and transfer and the different institutional actors within those sub-systems.

Along with these uncertainties, initial proposals gave NGOs a primarily implementational role – to transfer IBTA's technologies. None the less, some NGOs and their networks, such as PROCADE, have sought to negotiate the proposals in order to increase NGO influence over IBTA. Indeed, since the initial proposal in 1989, various NGOs have participated in activities with this goal in mind.

PROCADE believes that progress has been made. PROCADE is now a member of IBTA's national level Council of Directors. Similarly as the process of planning research activities has begun, PROCADE and other NGOs have participated in a series of departmental-level meetings in which technological priorities were identified and research plans drawn up. At the same time, as relations have improved, PROCADE has taken more initiatives to bring IBTA and PROCADE's member NGOs closer together in meetings to discuss research and collaboration. PROCADE also prepared a directory of its member institutions, listing their addresses, lines of action in ATD and their agronomic staff. This was to be made available to IBTA researchers (among others) in order to help them make contact with potential NGO collaborators working in the region of the IBTA research stations.

Sources: Bojanic 1991; Gonzalez 1991; Miranda, personal communication, 1991.

inter-institutional relationships were formalized between CIAT and NGOs in which CIAT conducted activities in the applied research sub-systems, shared activities with NGOs in the adaptive sub-system (and some in the applied sub-system), and left transfer activities to NGOs. These arrangements were, however, arrived at through a series of discussions between CIAT, donors and NGO staff. These discussions grew out of a strong network of informal relationships between individuals in these different institutions who knew each other's work methods, institutional circumstances and approaches to agricultural development. Some individuals have moved jobs between NGOs and CIAT. In certain cases political party affiliations also cut across these institutional lines. This mutual knowledge, and the relative lack of mutual suspicion, helped CIAT accept the idea that NGOs should have an influence over certain research decisions at CIAT (Box 6.13 and 6.14).

Box 6.13 CIAT creates space for NGOs

In 1983 CIAT ceased to do extension work as a result of the decision by the Departmental Development Corporation (CORDECRUZ) to conduct public-sector extension operations. CIAT concentrated its resources on research, coordinating with CORDECRUZ for the transfer of its technologies. However, the extension unit in CORDECRUZ was closed in 1987 as a result of rapidly falling revenue, leaving CIAT without a formal outlet for technology transfer.

None the less, it soon became clear that CIAT researchers had a number of informal contacts with NGO, producer, government and commercial organizations and that these contacts were serving as conduits for the transfer of CIAT technology, the contacts themselves channelling the technological information to farmers. In an internal research review of this situation, these contacts became known as 'intermediate users' (IUs) in CIAT lexicon, and it became clear that the closest and most effective contacts were with NGOs.

CIAT hosted a workshop with NGOs and other IUs to discuss efforts to begin organizing these linkage mechanisms with more formality and efficiency. Out of these discussions emerged a number of relationships and mechanisms which not only provided for the functional division of tasks between CIAT and NGOs but also created mechanisms whereby NGOs were involved in setting CIAT's research agenda for work with small farmers.

Sources: Veléz and Thiele 1991; Thiele *et al.* 1988.

Collaboration between NGOs and CIAT has also been favoured by a number of other factors (Box 6.14). First, NGOs openly admit their need for technical support (Garcia *et al.* 1991). Working in a zone of recent colonization, they are not always sure of the most appropriate technologies for sustainable resource management.[7] Consequently NGOs perceive CIAT's accumulated research results as an important resource. This was particularly the case in agroforestry, an area in which NGOs lacked experience but in which CIAT had a specially funded project. Moreover, the relatively high levels of funding at CIAT in the past mean it has been able to sustain work of a quality that has commanded the respect of NGOs and others.[8] Third, there is overlap between NGOs' and CIAT's perspectives on appropriate technology – CIAT has for some time incorporated systems, agroecological and agroforestry approaches into its work, and many NGOs see a key role for modern technologies and mechanization in solving the problems of colonist farming. Finally, NGOs have been encouraged by the special institutional arrangements in which CIAT is administered by a Board of Directors drawn from representatives of local organizations. This form of administration buffers CIAT at least in part from the influence of national political manipulation and of government in general.

In short, the functional and institutional reorganization of ATD at CIAT

Box 6.14 Factors favouring CIAT's relations with NGOs in Santa Cruz

- Social factors:
  - history of less political violence in Santa Cruz
  - strong personal relationships between CIAT and NGO staff
  - movement of staff between NGOs and CIAT
- Institutional factors:
  - CIAT was strong and donor funded in key areas (e.g. agroforestry research)
  - CIAT was partly protected from political interference
- Technological factors:
  - CIAT and NGOs share opinions on the 'appropriate technology' for poor farmers
  - NGOs recognize the need for technology research
  - strategic gaps in farmer technical knowledge

evolved steadily on the basis of local experience and existing inter-institutional relationships (unlike the changes in IBTA which were imposed quite abruptly by external agents responding to a rapidly deteriorating economic context). Overall, the context in Santa Cruz has been far more propitious for successful NGO–public-sector relations than has been the case in the Bolivian highlands. In certain respects the context in Santa Cruz was similar to the favourable context in Chile (informal contacts between NGOs and NARS, a strong public sector, overlap in NGO and NARS approaches to development etc.).

However, contextual factors change, and it is important to recognize that the quality of NGO–CIAT relations is not cast in stone. Indeed, in the last few years, political interference in CIAT has increased. At the same time, CIAT's financial situation has become more acutely difficult due to the declining tax revenues on which CIAT depends (these are channelled to CIAT via the Departmental Development Corporation). These two conditions are leading to weaker institutional autonomy and research capacity, which in turn may damage the quality of NGO–CIAT relationships.

Colombia

In Colombia the opportunities opened to NGOs by public sector and NARS reforms remain more potential than actual. Unlike INIA (Chile) or IBTA (Bolivia), the changes in ICA, the Colombian NARS, have made little explicit reference to NGOs; they do, however, create the possibility that NGOs will be able to exploit them as a means of gaining access to public-sector resources and decision-making.

The changes in ICA, and the overall reorganization of ATD in Colombia, are a combined consequence of constitutional changes that have sought to broaden the base of electoral and representative processes, and of efforts to reduce the expenditure of central government (Estrada 1991; Ritchey-Vance 1991; Wilson 1991). Central to these changes has been a programme of municipal reform and administrative decentralization. As part of this decentralization, ICA will phase out its own extension service, and municipalities will have to establish their own Technical Assistance Units (TAUs) to provide extension services by 1992. Central government will cease to finance this local extension, and these TAUs will have to be funded entirely from local tax revenues (Wilson 1991: 17).

The objective is to make technology transfer the responsibility of other public, parastatal and private agencies. ICA will provide the formal training that private extension groups will be required to receive before they are eligible to give extension in the TAUs (Wilson 1991). These provisions give NGOs an opportunity to do contracted work for municipal government and thus to gain access to the technical expertise of ICA.

While such contractual work will not allow NGOs any influence over local agricultural development policy, another related change does offer this possibility: the creation of Committees for Agricultural Technology Transfer and Technical Assistance within the Regional Councils for Agricultural Development (Consejos Regionales de Desarrollo Agropecuario). These committees, whose role is to elaborate departmental strategies for the coordination of technology-transfer work, will be constituted by public and private entities. In mid-1992, FUNDAEC and one other NGO, began participating in the Committee for the Department of Valle del Cauca. It is clearly too early to assess what impact this will have.

Ecuador

We mentioned in Chapter 3 that one form that the privatization of research and extension has taken has been the creation of agricultural development foundations. In 1988 the creation of one such Foundation in Ecuador, FUNDAGRO, marked an important stage in the restructuring of both the organization and the financing of the countries' research and extension services.

To create the Foundation, USAID diverted to FUNDAGRO resources for ATD projects which it had previously channelled to the public sector. In addition, it contributed an endowment, the interest on which was intended to cover FUNDAGRO's core costs. FUNDAGRO then sought additional support from other donor agencies for additional ATD projects. With these funds, FUNDAGRO implements activities in conjunction with public institutions, especially INIAP. Frequently INIAP will cover the staff wages of its researchers' involvement in projects, and FUNDAGRO will finance operating costs. FUNDAGRO, however, retains ultimate responsibility for monitoring, evaluation and final decision-making.

The justifications for the creation of FUNDAGRO were all related to failures of the public sector in ATD actions: its failure to coordinate research and extension, its growing inability to cover the recurrent costs of research and extension actions, and its politicization. FUNDAGRO was intended to be a mechanism that would improve administrative efficiency and would channel top-up funds to the public sector where needed (Coutu and O'Donnell 1991). What the creation of a professional, well-staffed foundation also meant, however, was that FUNDAGRO began to divert resources that otherwise might have gone to INIAP. At the same time, it captured several highly trained professionals from the public sector, attracted by the work conditions and higher wages that FUNDAGRO offered.

Initially many NGOs criticized FUNDAGRO. They saw it as an attempt to privatize agricultural research, which they felt should be a public service. None the less, its creation will have important implications for them and their relationship with the state. While it has not yet channelled resources through NGOs, this may happen in due course. More importantly, though, is that the creation of FUNDAGRO has led to a counter-response in INIAP, encouraging it to address some of the failings that justified FUNDAGRO in the first place. INIAP has sought, and in 1992 was granted, what has been termed 'institutional autonomy.' This will allow it independence from the political interference of government, and the freedom to set its own wage structure and raise its own funds. In return for this, over time INIAP will receive decreasing levels of funds from the state.

For NGOs the implications of these changes in INIAP are ambiguous. Assuming INIAP begins to attract higher-quality professionals, it will begin to address the problems of deteriorating institutional quality long criticized by NGOs. On the other hand, an increasing commercial orientation may reduce INIAP's willingness to provide technical support to NGOs free of charge. This may mean that NGOs will have to begin contracting research support from INIAP, making collaboration more expensive than before. However, freed from the trammels of public bureaucracy, INIAP ought in the future to be more flexible in negotiations, and certainly less distorted by political interference.

Aside from what these changes may imply for greater NGO contracting of INIAP services, other trends may lead to greater operational collaboration between the NARS and NGOs. INIAP will still receive donor and (in the mid-term) public monies to implement ATD activities. As it too looks to reduce its staffing and gain greater impact it will look to linkages with NGOs in programmes of adaptive research, technology validation and transfer (Cardoso 1991; Mestanza, Director of INIAP, personal communication, 1991). In some such cases, the NGOs' services may be contracted by INIAP or the funding agency. In PRONADER, for example, the rural development programme will channel funds to INIAP and to NGOs for local development actions. In this case, NGOs will be on contract to PRONADER and will be expected to collaborate with INIAP.

Peru

During the research, people were not keen to make any prediction about tendencies in the Peruvian NARS and other ATD institutions. The profound economic and political upheavals mean that there are many uncertainties about future policies. The trend, however, is clearly towards privatization.

As in Ecuador, an agricultural development foundation, FUNDEAGRO, was created in Peru under USAID auspices in 1987. According to commentators, it has devoted its efforts to the stimulation of private-sector companies for tasks such as seed multiplication and extension for instance (IFAD 1991; Coutu and O'Donnell 1991; Byrnes 1992).

The post-1990 Peruvian government has subsequently endorsed such private-sector activity, and has also passed legislation under which research stations, extension and special regional projects have become the responsibility of regional governments (IFAD 1991). Lacking funds these in turn look to private institutions and NGOs for support and for resources to help finance research station activities. The staffing levels of public research and extension have also been drastically reduced (IFAD 1991).

The collapse of the Peruvian public sector is the most dramatic of the five countries. In many areas of the country, any technical assistance that is given now comes from NGOs (in many others, civil war has meant no assistance is given at all). Likewise, private and non-governmental support is increasingly necessary to rescue the infrastructure that the public sector has developed in the past, but which it is now unable to maintain – such as agricultural research stations.

EMERGING PATTERNS OF REFORM IN THE ATS: OPPORTUNITIES AND CHALLENGES FOR NGOs

Functional linkages, institutional roles and contextual factors

While there are clear differences among the reform programmes discussed above, there is a common pattern. The neo-liberal tide in the region has led to efforts to reduce government expenditure and involvement in ATD activities – particularly in technology adjustment and transfer activities. Government is pulling away from the implementation of agricultural and rural development activities, and research institutes are looking to become generators of technologies that they will then pass on to 'intermediate users'. The 'intermediate users' are local public agencies, the private sector, NGOs and producer organizations. In some cases, such as Peru and Ecuador, they also hope these users will help contribute to the costs of research.

In functional terms these changes might allow the most efficient division of tasks in different sub-systems of the ATD among NGOs and NARS. NARS would be primarily responsible for activities in the basic and applied research sub-systems; NGOs would share tasks in the adaptive sub-system with NARS;

and NGOs (along with private technology-transfer companies and farmer organizations) would be the main actors in the transfer sub-system, simultaneously pursuing other community development activities to facilitate farmer incorporation of new technological options and to promote other dimensions of social development.

The operational effectiveness of the links between the actors in the sub-systems will be one factor in determining how well this chain of functions is integrated. These linkages are discussed in the next chapter. However, contextual factors will also influence the quality of the relationship between governmental and non-governmental actors in different parts of the ATS (Box 6.15). These contextual factors remind us that we cannot analyse NGO–NARS relations only in functional terms, and that if we do so our analysis may be naively over-optimistic about the scope for NGO–NARS coordination.

Box 6.15 Contextual factors influencing the impacts of public-sector reform on NGO–NARS relationships

Contextual factors that favour closer NGO–NARS linkages

- Informal contacts between NARS and NGO staff.
- Pressure from donors for collaboration.
- Shared vision of 'appropriate technology'.
- Neither sees each other as a pure instrument to serve their own ends.
- All participants involved in development of proposals for NGO–NARS collaboration.
- Both NARS and NGO are skilled and committed to quality.
- Return to electoral democracy

Contextual factors frustrating closer NGO–NARS linkages

- Government commitment to socially regressive structural adjustment programmes.
- Lack of informal contacts and mutual understanding.
- Imposition of collaboration from outside.
- Weak NARS technical capacity due to persistent underfunding.
- Exclusion of NGOs from decision-making processes in NARS planning.

Of course, some of these factors, notably the return to electoral democracy, have changed so as to facilitate the NGO–NARS relationship. Equally, the general context of reorganizing government institutions and the interest in NGOs has helped create an environment in which government institutions and NGOs are having greater contact with each other. Consequently they are revising some of their opinions and distrust on the basis of greater mutual knowledge.

Other changes in context, however, have created problems. Most difficult is governmental endorsement of structural-adjustment programmes that NGOs and many farmer organizations perceive as having socially regressive and unacceptable impacts. This disagreement over economic policy, even though not directly related to ATD, can none the less affect these relations within the ATS.

In addition, NGOs question the continuing tendency of NARS to separate technology from national and local socioeconomic context. They argue instead that technology development work should be combined with efforts to change policies and institutions in areas such as land tenure, credit and marketing. The scope for broaching such wider changes in their discussions with government will influence the quality of, and NGO commitment to, more functional relationships purely concerned with technology issues.

For the system to function well, and to generate 'appropriate' technologies, actors in the downstream sub-systems must be able to exercise influence over actors 'upstream.' This so-called 'feedback,' whose success depends to a great extent on farmers and extensionists having means through which they can influence actors in research sub-systems, has historically been the weak point of NARS. The problems faced by the PIPs in Ecuador in being able to gain credibility among INIAP researchers was just one illustration of this. It is not clear is whether, under current changes, NGOs will have any more success in establishing this authority than the PIPs did.

One way in which NGOs might be able to exercise such influence is by contracting research from NARS. Assuming that NARS continue to put greater emphasis on raising their own funds, this option will be increasingly open to NGOs. Of course, it is hardly the sort of relationship they had expected to enter with NARS, and many will find it difficult to accept. However, it will offer one powerful way of holding NARS to account and directly affecting their agendas. To those donors concerned that NARS be more campesino oriented, giving NGOs funds specifically earmarked for commissioning research from the NARS may be an interesting option.

Reforming the ATS: new challenges to NGOs

These ongoing changes within the NARS and its relations with other actors in the ATS imply a range of new challenges for NGOs. The first challenge is how to negotiate these emerging relationships in a way that will allow operational contacts at the same time as it increases the NGOs' capacity to influence NARS decision-making. Here, informal contacts and donor leverage in support of NGO concerns will be important contributory factors.

A second set of challenges for NGOs comes from the increasing pressure they face not simply to have contact with government but to implement government funded or coordinated contracts (e.g. INDAP in Chile, and PRONADER in Ecuador) or to assume coordinated roles with government in project activities. As they assume these roles, NGOs begin implementing parts of wider programmes managed by public agencies. This raises several challenges for NGOs (Aguirre and Namdar-Irani 1991; Sotomayor 1991):

- to their *financial security*: the more that NGOs accept government funded contracts, the more they are obviously vulnerable to the winds of change in government.
- to their *autonomy*: as they commit themselves to coordinated programmes, NGOs surrender a certain degree of autonomy over their own actions and expose themselves to external factors that might disrupt such programmes. Clearly this is the more so if NGOs are at the same time accepting government funding.
- to their *cohesion*: as NGOs commit themselves to the mandates of other state institutions, then this can create tensions in the NGO. Whether these will be transitory or destructive will depend on whether the NGO is able to change the mandate and approach of the public institution, before the mandate changes the NGO (Box 6.16).
- to their *identity*: NGOs' prior identity was previously grounded in a critique of government, and in opposition to it. As changes in the NARS proceed to take on board some of these critiques, and government programmes begin to approach NGOs for advice and assistance, so NGOs are challenged to rethink their identity: no longer can it be an identity purely of opposition.

One of the case-study authors suggested that in the long term, for progressive NGOs to retain an institutional identity that is consistent with their political position then the more viable strategy for them is one focused on developing innovations for presentation to government, rather than on implementational work. The more they commit themselves to the latter the more they may lose their identity and motivating vision. They will become increasingly the implementers of somebody else's programmes (Sotomayor 1991). However, while this may undermine the identity of those NGOs formed under difficult political circumstances, it may be no problem for those 'opportunistic NGOs' created much more recently, and whose reason for existing is largely to implement programmes rather than to effect any social change.

A third challenge to NGOs is to upgrade their professional standards. As more attention is paid to their work, and more of the burden of adaptive research and transfer is placed on them, many recognize the importance of improving the technical and economic quality of their work. While their prior roles of policy critic, popular educator and grassroots educator benefited from the skills of anthropologists and sociologists, their new roles as programme implementers and policy advisers (as opposed to critics) will require far higher levels of expertise in, for instance, agronomy, economics, animal sciences and small-business administration.

A final challenge, so far limited to Chilean NGOs, is that which arises where a political change leads the NARS and other government institutions to begin assuming financial or implementational roles and responsibilities in rural

Box 6.16 When innovation in the public sector forces NGOs to innovate

The decision since 1990 of the Chilean public sector to increase its involvement in the development of peasant agriculture has had important consequences for AGRARIA's own strategy as an NGO. The most significant are the following:

- AGRARIA continues to implement technical assistance programmes in the areas in which it traditionally worked, but has in many cases gained INDAP funding for these. Although this has introduced a rigidity in field methods, AGRARIA believes it is no longer viable to consider delivering NGO funded technical assistance now that state resources are available.
- AGRARIA has refocused its available resources towards activities in which the state is not working, such as agro-industry, the development of producers' organizations, marketing, irrigation, afforestation and resource-conservation. This has made it necessary to retrain professional teams in these fields. It also implies a shift from solely implementing programmes – as in the case of extension – to formulating them for presentation to the public sector and foreign agencies for wider replication.
- AGRARIA has also dedicated more effort to participating in the debate with INDAP on appropriate development and extension strategies for peasant agriculture. However, it is increasingly difficult to obtain funding for this type of analytical work. Therefore, AGRARIA has initiated a strategy to generate funds for this type of work by selling its services to INDAP and FOSIS.

Source: Aguirre and Namdar-Irani 1992.

development that were traditionally the preserve of NGOs. How far the Chilean experience may point to challenges that will arise for NGOs in other countries is debatable. Chile differs in that the public sector has resources with which to finance ATD work in ways that the other countries do not, with the possible exception of Colombia. In Chile the Aylwin government has made a commitment to rural development and launched a relatively massive programme of technical assistance to the rural poor financed through INDAP and, to a lesser extent, the social investment fund, FOSIS.[9] In doing so, the government has assumed financial responsibility for areas of activity previously dominated by rural NGOs. These NGOs have therefore been presented with a twofold problem. On the one hand, there is a squeeze on financial resources. The government has created an agency specially mandated to capture donor funds for poverty alleviation, and in so doing is diverting bilateral money that was previously channelled through NGOs when they were the only institutions working in poverty alleviation. On the other hand, it now becomes increasingly difficult for NGOs to justify continuing the type of technical assistance work upon which they traditionally concentrated, because they would be merely spending money in projects paralleling state actions, and would thus be using donor money for actions that they could fund with contracts from the state. This pushes NGOs into a double challenge: to identify a new role for themselves (Box 6.16) and to tackle the problem of how to cope with contracts from the public sector (Box 6.17).

Box 6.17 The problems of collaborating with government: a Chilean experience

The Chilean NGO, AGRARIA, made an institutional decision to accept a large number of INDAP contracts as a potential opportunity to influence the use of government funds for rural development and poverty alleviation. None the less, in doing so, it has had to take on many new staff, has almost doubled in staff numbers and has had to adapt those parts of it working on INDAP contracts to the contractual requirements of INDAP: some of these require a field methodology that is at odds with the traditional methods of AGRARIA. These factors have introduced imbalances in the NGO:

- The new staff are not always as ideologically committed, nor identified with the institution's own political trajectory, as the older staff.
- The organization has grown so large (over 100 staff) that it is now impossible for all members of AGRARIA to know each other and identify with its other field programmes.
- In addition, there is the threat of a division between the part of AGRARIA that acts as AGRARIA has always done, and those parts that even within AGRARIA are referred to as 'AGRARIA-INDAP' (this phenomenon has also occurred in GIA as it too has adopted INDAP contracts).

Source: Aguirre and Namdar-Irani 1992; Aguirre, personal communication, 1991; Sotomayor, personal communication, 1991.

CHAPTER SUMMARY

In this chapter we have argued that over recent years a series of changes in the non-governmental sector, and above all in the public sector, have increased the potential for more collaborative forms of interaction between the two. Elements of these changes, however, have derived from policy decisions, or had effects, that older more radical NGOs initially find it difficult to accept. This will complicate their response to these new opportunities.

The increased tendency to privatize certain aspects of ATD activities in the public sector is leading to a situation in which NARS are concentrating their resources in applied and adaptive research activities, and expecting other institutions to absorb some of the costs of implementing transfer activities. In many cases, the NARS are also hoping to gain access to resources currently controlled by NGOs, in order to supplement the scarce resources in the public sector for applied and adaptive research.

In most cases the NARS have gone around implementing these changes in a rather top–down way and have mistakenly assumed that NGOs will simply fall in line with what NARS want them to do. This attitude has annoyed many NGOs, making them feel excluded from decision-making. It has therefore frustrated early discussions on future collaboration in ATD. Nevertheless, the overall weakness of the public sector means that it is often in no position to dictate to NGOs. This gives NGOs a degree of power to renegotiate some of these changes. Likewise, even if NGOs dislike the idea of sharing resources and perhaps even paying for parts of NARS research, these changes will increase the scope that NGOs have to influence how public resources are used. The changes have, then, created more room for manoeuvre for NGOs: in their own way, public-sector reforms have undoubtedly had impacts that are potentially democratizing.

These gains do not, though, come free. The costs to NGOs are that these new relations will threaten their identity, cohesion, autonomy and in some cases financial independence. These threats demand creative responses from NGOs. They do not, though, mean that NGOs should reject these new opportunities. Indeed, the bottom line for those more radical and progressive NGOs on which we have focused our attention, is that they have little choice but to respond creatively. If they do not, the more opportunistic organizations and the private sector will occupy these new relations with government, gain influence over decision- and policy-making, and ultimately grow to the extent that they marginalize other NGOs who do not grasp the nettle.

The question of linkage is thus now irredeemably on the agenda: both as a means of enhancing the efficiency of ATD through improving the quality of links between actions in different sub-systems, and as a means for negotiating these new relations. How to manage these linkages to best effect is thus a central question to NGOs and NARS alike. In the next chapter we review past experiences of linkages in order to offer reflections on future possibilities.

NOTES

1 At the workshop in Santa Cruz to discuss this research it was recognized that 'the greatest cause of NGO project failure resulted from the weak coordination with other institutions, other NGOs and the public sector present in the area in which the NGO was working' (Bebbington *et al.* 1992: 13).
2 These courses have frequently been given at the instigation either of donors who wish to see a change in NARS actions, or of personal contacts of NGO staff within the NARS.
3 The terms 'network' and 'consortium' are used quite loosely here. Strictly speaking they have different functions. Our main intention in referring to them in this section is to refer to efforts to increase and formalize coordination and contact among NGOs.
4 The enthusiasm for networks is reflected in the following comment made at a meeting of Bolivian networks in 1989: 'The development of national level networks offers the possibility of much more effectiveness in the coordination of projects and programmes, in the recovery of NGO experiences and the dissemination of these experiences on a far wider scale' (Anon. 1989: 5).
5 UNITAS is the Unión de Instituciónes para el Trabajo de Acción Social (Union of Institutions for Social Action), and AIPE is the Associacion de Instituciónes para la Promoción y Educación (Association of Social Development and Education Institutions).
6 Andean crops programmes focus on crops such as *quinoa*, *tarwhi*, *cañihua*, *oca*, *melloco* and *mashua* – all indigenous grains, legumes and tubers.
7 In such areas of rapid change, farmer knowledge does not offer the same clues as it does in areas of more stable population (such as the Andean region).
8 This is in part a fortunate effect of higher tax revenues in Santa Cruz due to fossil fuel

deposits in the department. CIAT's funding has also been boosted from the ODA-funded British Tropical Agricultural Mission.

9 According to World Bank sources (the funding agency for INDAP), this initiative has been conceived of as a poverty alleviation programme as much as an agricultural development strategy (Wilson, personal communication, 1992).

7

MAKING NGO–NARS RELATIONS WORK

The evidence on linkage mechanisms

Chapter 6 looked at changes in the structure of the ATS, changes which – particularly the radical restructuring and retrenchment of the NARS – are forcing the issue of NGO–NARS links up the agenda. We also considered some of the ways in which NGOs may try and renegotiate such links, or indeed use them to exercise influence over the public sector. In this chapter we concentrate more on the functional dimension of these links. We consider what sorts of linkage mechanisms might be used in order to allow more efficient two-way flows of technical information between institutions, and to strengthen the contributions of NARS and NGOs to the generation of technologies that are relevant to campesinos. The chapter also provides a framework for the analysis of linkage mechanisms between NARS and NGOs. The experiences of the case studies are considered within this framework.

We begin the chapter looking at different functions that may be performed by the NARS and the NGOs in the development of agricultural technology (ATD). The different ways in which these functions can be divided among actors may be described as 'technology paths'. We then examine linkage mechanisms in relation to different technology paths. The chapter finishes with a discussion of critical issues to be addressed by managers of NARS and NGOs when considering linkages. We emphasize, following Merrill-Sands and Kaimowitz (1990), the need to make linkage mechanisms explicit and to develop them so they fit the specific goals of the institutions involved.

TECHNOLOGY PATHS AND MULTIPLE ACTORS IN ATD

Our analysis focuses on the functions performed by NARS and NGOs in three sub-systems of the ATS: applied research, adaptive research and technology transfer. There are many different ways in which activities for these functions can be divided between NARS and NGOs.

The dominant model for ATD in Latin America has been the public-sector National Research Institute, in which a single institutional actor is responsible for the functions of all three sub-systems. A variant of this model, such as that which has been implemented in Chile, assigns research and transfer to two different public-sector organizations. Another model proposed more recently, for example by IBTA, Bolivia, reserves the two research sub-systems for a single public-sector institution, but uses NGOs for technology transfer.

Against this more orthodox focus on public institutes, Biggs (1990) argues that a model of multiple sources of innovation provides a better explanation of the range of actors involved and ways in which technologies are developed. According to this model there exist many sources of technological innovation outside the public sector. These include not only NGOs but also farmers and private companies. If there are multiple sources of innovation, this implies there will also be multiple forms in which innovations are developed and transferred. These can be identified by following the route, or *technology path*, through which a technology, or group of technologies, is created, developed and transferred within the ATS.

Figure 7.1 shows a simplified matrix of technology paths which involve NARS and the NGOs. (Farmers do not appear in this matrix, although they of course engage in ATD activities: Rhoades and Bebbington 1993.) These paths are not exclusive. The same technology, or variants of it, might follow more than one path. If we were to include other agencies (private companies or farmers), or to consider situations where the two actors share the same function, the matrix would become much bigger. In spite of the simplifications, the matrix provides an idea of the range of possibilities for dividing functions between the two actors. Only the three sub-systems at the centre of the ATS (see Fig. 2.1) have been considered. Even so, in this simplified scheme there are eight possible different paths.

The matrix can be used to analyse different models or strategies for ATD. The first path in Figure 7.1 represents the dominant model of ATD in Latin America, where the NARS fulfils all three functions. The second path represents the more recent model where NGOs are given the task of technology transfer but their role is restricted to this.

Table 7.1 shows which of the eight different possible technology paths identified in Figure 7.1 were followed in practice by the NGOs which participated in the study. The table also shows the crop or cropping system, and the sub-system involved. Paths 3 and 8, where the NGO carries out a research function, were by far the most frequent. Neither the first model described above, where the NARS realizes all ATD functions, nor the second model, where the NGO is restricted to technology transfer, recognize the potential research roles that NGOs can play.

It is worth analysing in more detail the different paths which appear in Table 7.1 and the factors which determine which path was dominant in a particular case. All the technology paths we considered involved NGO participation, so no examples of Path 1 appear in the table. There were four cases of Path 2, where the NGO transferred a technology which had already been investigated and adapted by the NARS. This compared with eight cases of Path 3, where the NGO carried out adaptive research with a technology

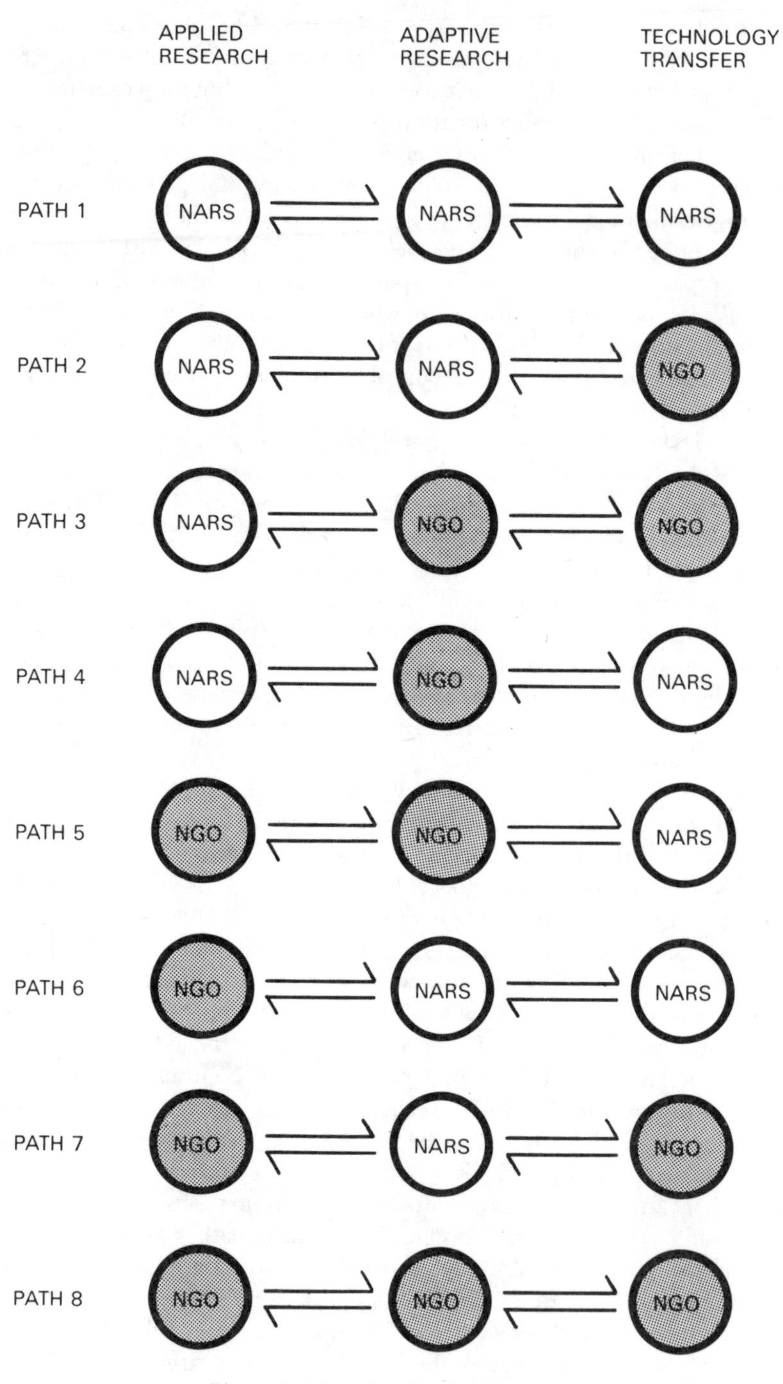

Figure 7.1 Matrix of technology paths

Table 7.1 Technology paths predominating in the case-study NGOs[a]

NGO	*Path 2*	*Path 3*	*Sub-function of Path 3*	*Path 4*	*Path 5*	*Path 8*
CESA-EC	Potato	Potato	Selection and adjustment			
CAAP						Andean crops
FUNDAEC				Cropping sub-systems		
SEPAS		Blackberry, irrigation	Adjustment			
IDEAS						Agroecological production systems[b]
CIED						Agroecological production systems[b]
CEIBO		Cocoa	Selection and adjustment			Cacao
CIPCA	Maize	Rice	Selection and adjustment			
CESA-BOL		Agroforestry	Adjustment			
Kohl (Bolivian NGOs)		Protected horticultural systems	Adjustment			
GIA	System Beans and maize	System Beans and maize	Adjustment		System Beans and maize	
AGRARIA	Pastures Wheat	Pastures Wheat	Selection and adjustment			

[a] See Figure 7.1.

[b] Whilst there was no NARS involvement in the development of the agroecological production system technology of IDEAS and CIED, there was university support for the research involved.

developed by the NARS (Box 7.1). Additionally, in all four cases of Path 2, the same NGO has had other experiences in which functions had been divided as in Path 3. It would therefore seem that Path 3 is particularly characteristic of NGOs' work in the development of agricultural technology.

Box 7.1 CESA-Ecuador, an example of Path 3: the NGO tests and adapts NARS technology

CESA seeks to find a middle way between ethno-development and Green Revolution strategies, represented in Ecuador by the CAAP and INIAP respectively. CESA believes that the strategy proposed by those committed to ethno-development to recover traditional agricultural technologies remains unproven and idealistic. CESA starts from the actual practices of peasant farmers, who have in many cases already adopted agrochemicals and 'modern' varieties. While seeing a role for these new inputs, CESA's programme of experimentation and demonstration tries to reduce the intensity of their use. This programme involves adaptive research, adjustment and diffusion of technologies. CESA tries to adapt INIAP's recommendations to the situation of the peasant farmer. The programme is carried out in ten different zones, with 150 grassroots peasant organizations and more than 8,000 families.

Source: CESA 1991.

The predominance of Path 3 can largely be explained by the fact that the NARS generally develop technologies for large farmers in resource-rich areas, whilst the NGOs work with small farmers in marginal areas. Consequently, some adaptive research is almost always necessary – indeed much of what passes in NGOs as field demonstration of technologies is in fact adaptive research. However, a second factor for the prominence of Path 3 is related to our deliberate selection of stronger, better endowed NGOs for the case studies – such NGOs are more likely to have carried out research.

In considering the type of adaptive research that NGOs conduct in these technology paths, it is helpful to break down that research into two different phases, or sub-functions.

- Phase 1: Technology selection. This involves the adaptation of the technology to an agroecological zone. Although this can be done on farm, it is frequently carried out in an experimental station, because of the greater control the station permits. Generally, research of this type has a complex design and is analysed statistically.
- Phase 2: Technology adjustment. This involves the adaptation of the technology to the socioeconomic conditions of the farmer. This type of research is carried out on-farm and with the participation, in varying degrees, of the farmer. The experimental design is simple and socio-economic evaluation is more important than statistical analysis.

The existence of these two phases in adaptive research further complicates the matrix of paths. The division of these two sub-functions between NARS and NGOs is examined, in relation to Path 3, in the third column of Table 7.1. It can be seen that NGOs carry out technology adjustment more frequently but that they have also performed technology selection.

Both AGRARIA (Chile) and CIPCA (Bolivia) initially carried out technology selection, but consciously withdrew from this phase when they realized that the NARS had the resources to do it better. This allowed them to concentrate on technology adjustment (Box 7.2). This modification in research strategies by the two NGOs also shows their flexibility in changing from one path to another when implementing a research programme.

Box 7.2 AGRARIA-Chile: from selection to adjustment of technology

Adaptive research carried out by AGRARIA in the Sauzal project has passed through three stages.

In 1984 investigation sought to identify technical factors to produce maximum yields of wheat, lentils, oats, chickpeas and sorghum. A complex multi-factorial design was used for experiments, in an attempt to reproduce the techniques of an experimental station. This approach was adopted because of the total absence of relations with INIA, the fruit of mutual mistrust given the political situation, and the orientation of the NARS towards large-scale farming. AGRARIA soon concluded that its limited resources, the lack of training of its technical teams and the availability elsewhere of information about yield-maximizing technologies limited the usefulness of this type of research.

In 1986 AGRARIA changed its strategy to one of adjusting technologies already developed in INIA's experimental stations. Instead of seeking yield-maximizing technologies, the NGO began to seek 'optimums' for campesinos; that is, they began to seek the technologies that would perform best within the campesinos context. For example, research on wheat used much lower doses of fertilizer which were more appropriate to farmer conditions.

In 1988 AGRARIA, continuing the strategy based on technology adjustment, introduced a systems focus. An informal research proposal was drawn up, together with specialists from INIA. In this stage the still informal relationship between certain scientists at INIA and the Sauzal project was one of reciprocal support. A relatively continuous flow of technical information from the experimental station to AGRARIA was established.

AGRARIA in turn, through its experimental work under farmer conditions, passed results to INIA's specialists. AGRARIA contended that it was necessary to increase the productivity of systems rather than work with isolated components. It succeeded in demonstrating to the INIA researchers the viability of a system based on rotations of wheat and a naturally occurring legume forage *hualputra*.

Source: Aguirre and Namdar-Irani 1991.

However, the NGO can restrict itself to technology adjustment only if the technologies being offered by the NARS correspond sufficiently to those needed by their client farmers. If this is not the case, then the NGO has two options: to engage in technology selection, or consciously to change the campesinos' circumstances, essentially adapting the farmer to the technology (e.g. with infrastructure support). Indeed, CIPCA's development strategy was to adapt campesinos' conditions so that they could make use of new technologies available from CIAT, by introducing farm mechanization into campesino production (see Annex 2, Profile 3).

In Path 4 the NGO carries out adaptive research on a technology generated by the NARS which subsequently takes charge of transfer. In Path 5 the NGO does both adaptive and applied research and the NARS takes responsibility for transfer. The cases of FUNDAEC (Box 7.3) and GIA were the only examples encountered of these two paths. These paths are not common, at least in part because information flows about research results from NGOs to NARS are weak. Path 4 had two advantages for FUNDAEC. It allowed FUNDAEC to make use of its own field of knowledge to adapt the technology to farmers' conditions, and it gave FUNDAEC access to the NARS technology transfer infrastructure which allowed the widescale diffusion of FUNDAEC's research findings – a form of scaling up the impact of the NGOs' research.

Box 7.3 FUNDAEC: an NGO carries out technology adjustment prior to diffusion by the NARS

In the north of the Cauca Department of Colombia, FUNDAEC has carried out research into alternative production systems for campesinos since 1973. This programme has identified and evaluated on-farm ten different production sub-systems. At present FUNDAEC is seeking to pass on its findings to public-sector institutions which work with a far larger number of farmers than does the NGO. As a first step, the NGO has organized courses for public-sector technology-transfer staff, including those of ICA and other public-sector institutions, to pass on information about its research results.

In 1989 FUNDAEC also entered a joint programme for the improvement of farming systems in the Mondomo region of Cauca. This involved ICA, CIAT and other local institutions. FUNDAEC's role was to train extensionists and farmers, using the results of both its own research and that of CIAT. Again, FUNDAEC is aiming to scale up its impact through training programmes to the public sector.

Source: FUNDAEC 1991.

The absence of any examples of Paths 6 and 7, in which the NGO carries out applied research and the NARS adaptive research, is hardly surprising. NGOs rarely carry out applied research, and when they do, they are under strong pressure to adapt the technology to the needs of their client farmers.

Four cases were found of Path 8. This path occurs when the NARS is effectively absent from the zone where the NGO works, or when the technologies the NARS offers do not correspond to NGOs' perception of farmers' needs. This latter situation is most frequent with agroecological and ethno-development NGOs. In Ecuador, for example, CAAP researched indigenous cropping practices, which were not covered by the NARS. The agroecological focus of CIED in Peru was completely different from that of the NARS, although there was collaboration with the university. In the case of El Ceibo, the NARS, because of financial constraints, was effectively absent from the area. Finally, in the case of NGOs working with protected horticultural systems, the NARS was not investigating the technology, which the NGOs considered to be important. Consequently, the NGOs had to carry out all the research functions.

Technology Path 8 represents the most radical model of an alternative organization of the ATS. It is the alternative pursued by those NGOs which consider the technologies being offered by the NARS to be completely inappropriate. It is a mirror image of the dominant model of ATD in Latin America, where the NARS carries out all of the functions in the ATS. It similarly fails to exploit possible complementarities between the two actors.

In Paths 2 to 7 there is a division of functions between NGOs and NARS, although this does not necessarily imply institutional contact between the actors involved. However, there are many interdependencies between actors performing different functions. This means that the efficient operation of the ATS requires coordination between these actors, and particularly between those mainly responsible for research and those mainly responsible for extension (Box 7.4). To achieve an efficient coordination requires the existence of linkage mechanisms, which are discussed in the next section.

Box 7.4 Interdependencies between actors carrying out research and technology transfer

Researchers need the support of technology transfer agents in:

- selecting the right technology for development
- appropriate design of trials
- identifying representative locations for trials
- providing feedback about farmers' reactions to new technologies
- interpretation of trial results

Technology transfer workers need the support of researchers in:

- providing information about the characteristics of new technologies
- explaining the context in which new technologies should be used
- ensuring that the messages being delivered to farmers are correct and sufficiently detailed
- providing specialist services such as soils testing and pest identification

Source: Merrill-Sands and Kaimowitz 1990: 16.

TYPES OF LINKAGE MECHANISM

Linkage mechanisms help the transmission of information about technologies between the NARS and NGOs and increase the possibility of coordinating activities. Frequently the institutions involved do not recognize the existence of linkage mechanisms, or pay little attention to their organization. As a result many linkage mechanisms are poorly structured or based on informal contacts between individuals. In certain cases, this may often be a more efficient arrangement (see below).

The nature of the linkage mechanism depends on the functions performed by each actor in the ATS, as well as a range of contextual factors. We have seen in the different case studies that the functions being performed by NGOs and NARS vary considerably, so it is no surprise to find a wide range of linkage mechanisms.

The ISNAR studies discussed in Chapter 2 found two main types of linkage mechanism (Merrill-Sands and Kaimowitz 1990). Adapting their terminology to include both NARS and NGOs as actors in ATD we have:

- *operational* linkage mechanisms which coordinate or support the implementation of specific ATD functions by the two actors;
- *structural* linkage mechanisms which function at a general planning level to create linkages between NARS and NGOs.

Operational linkage mechanisms

The ISNAR study identified four types of operational linkage mechanism and the same terminology can be used to analyse links between the NGOs and the NARS:

- planning and review processes;
- collaborative professional activities;
- resource-allocation procedures;
- communication devices.

The principal mechanisms identified in each category are shown in Box 7.5.

These operational mechanisms can be classified according to whether they are formal or informal, mandated or voluntary, permanent or temporary (Kaimowitz *et al.* 1990: 233). Formal linkage mechanisms are officially recognized by the institutions involved, whereas informal linkages are based on personal relations between individuals. When the institutions involved can impose their decisions at the level of the two institutions, the mechanisms are *mandated*. However, when they can only suggest or influence they are *voluntary*.

As has been discussed in earlier chapters, it is frequently the case that an NGO establishes a linkage mechanism for a specific activity because it wants to influence public-sector policy. It may seek to change the way the NARS

Box 7.5 Operational linkage mechanisms

Planning and review processes:

- Joint problem-diagnosis
- Joint priority-setting and planning exercises

Collaborative professional activities:

- Formal collaboration in trials, surveys and dissemination activities
- Joint decision-making on release of recommendations
- Regular joint field visits
- Informal consultations

Resource allocation procedures:

- Formal guidelines for allocating time for collaborative activities
- Specific allocation of funds for collaborative activities
- Staff rotation and secondment

Communication devices:

- Publications, audio-visual materials and reports
- Joint training activities or seminars

Source: Merrill-Sands and Kaimowitz 1990: 52.

allocates resources, sets its research agenda or carries out certain functions. In this chapter, however, we emphasize the functional aspects of these linkage mechanisms. Their use by NGOs for policy leverage with NARS is another important aspect. We have separated the political and the functional between chapters for analytical purposes. In reality the two aspects frequently occur simultaneously.

Structural linkage mechanisms

Three different types of structural linkage mechanism were found in the case studies:

- coordination units, which have the specific function of coordinating activities between the two actors;
- permanent committees made up of representatives of the NARS and the NGOs;
- representation by one actor in the managing body of the other.

Structural linkage mechanisms are necessarily formal. They are usually permanent because they seek a more sustainable coordination between institutions. They vary in their ability to impose decisions – some may be mandated, others voluntary.

EXPERIENCES WITH OPERATIONAL LINKAGE MECHANISMS

Table 7.2 shows the different operational linkage mechanisms which were found in the case studies, distinguishing those described by the NGOs and those by the NARS.

The first column of the table indicates the relative importance of the linkage mechanism for the institution for the success of its work in ATD as described in the case studies prepared.

Linkage mechanisms were important in only five of the fourteen cases and all of these were in Bolivia and Chile, where, as discussed in Chapter 6, structural linkage mechanisms have been introduced.

At the other end of the spectrum there are cases in which the NGO assigns no importance to linkage mechanisms. This is not surprising because the three NGOs in which this is the case (CAAP, IDEAS and CIED) are those which explicitly follow Path 8 and deem most NARS technology to be of little or no relevance to their work.

In the other seven cases where linkage mechanisms are given some importance, the institutions concerned participate in technology paths which involve NGOs and NARS. In these cases one might hypothesize that if the mechanisms were accorded more importance, and so planned and monitored with more care, then the coordination between NARS and NGOs might be more effective. This assertion was generally supported in the Santa Cruz workshop, although all involved recognized that one reason for the neglect of mechanisms was lack of time and resources to support them.

The second column of the table indicates whether formal or informal linkages are relatively more important. Formal linkages predominated in only four cases. All of these were in Bolivia or Chile, where structural linkage mechanisms have been created. This favours the formalization of operational linkage mechanisms. There were specific cases of formalization in other countries but these were rarer – indicative of the more recent changes in the NARS and, in the case of Peru, the rapid disintegration of the NARS (see Ch. 5).

Probably the most common informal linkage mechanism is the one-off consultation between individuals, which does not require complex coordination between the actors involved and requires no institutional participation. However, other mechanisms, which require more sustained and complex coordination, such as collaboration to carry out a trial, can also be carried out informally.

Informal linkage mechanisms are useful. They reduce paperwork and delays, cut through bureaucratic procedures and exploit personal contacts. However, as can be seen from the case of PIP-INIAP (see Box 7.6), they have their limitations. During the Santa Cruz workshop it was noted that personal or informal relations provide a strong base while the people involved remain in the institution. However, if there is a change in staff, the confidence which was built up is lost and the inter-institutional relationship ends (Bebbington *et al.* 1992).

The last four columns of the table correspond to the four different types of operational linkage mechanism. The specific linkage mechanisms identified in each case study have been briefly described in the appropriate column.

'Collaborative professional activities' is the most common type of linkage mechanism, which appears in twelve of the fourteen case studies. 'Planning and review' appears in half of the cases and 'Communication devices' in approximately a third. The mechanism 'Resource-allocation procedures' appears in only one case. Once again, the majority of these cases are to be found in Chile and Bolivia where structural linkage mechanisms exist.

At present it seems, then, the full range of linkage mechanisms is not being utilized. Furthermore, the fact that the principal mechanism is a rather loosely defined category labelled 'collaborative professional activities' is indicative of the fact that most linkages occur at a field level, between individuals and beyond the constraints and dictates of institutional procedures. For other mechanisms to be more important, they would have to be recognized and encouraged at an institutional level. This has been far rarer. So far, those who are most likely to recognize the benefit are those who spend most time in farmer's fields.

In the next section we consider the characteristics of the different operational linkage mechanisms in more detail.

Joint planning and review processes

These processes can be classified into four types:

- joint problem-diagnosis
- joint planning
- joint programming
- joint evaluation

Each of these will be considered individually.

Joint problem-diagnosis

The case studies revealed two examples of this type of mechanism in action. In Ecuador in a recent baseline study to design a new project, CESA organized a diagnosis activity in Chimborazo in which NARS staff participated informally. In Bolivia a NARS, CIAT, organized a diagnosis using the *sondeo* methodology, with the participation of several NGOs in Santa Cruz. Not only did this make use of the local knowledge of the NGOs, it also established the base for future NGO–NARS coordination in the zone (Box 7.7).

Table 7.2 Operational linkage mechanisms found in the study

	Importance	*Dominance of formal (F) or informal (Inf.) contacts*	*Planning and review*	*Collaborative professional activities*	*Resource-allocation procedures*	*Communication devices*
(1) **NGO cases**						
CESA-EC	**	F, Inf.	Committee in rural development programme combines both CESA and public institutions	– Technical advice from NARS for adaptive trials type I and II[a] – Coordination for adaptive trials type II – Coordination for transfer in fruit technology – General agreement for technical cooperation	NARS lent expert in fruit production	
CAAP	*	Inf.		– Assisted NARS in locating trials		
FUNDAEC	**	F, Inf.		– Collaborative livestock project for research and transfer		
SEPAS	**	F, Inf.				
IDEAS	*	Inf.		– Diffusion of technology		
CIED	*	Inf.				
CESA – Bol.	***	F		– Visits by NARS researchers to CESA – Study visits to NARS		– Agroecology symposium with NARS
CIPCA	***	F	– Participation in annual planning of NARS	– Jointly-run experimental station with NARS – NARS field days		– NARS bulletins – NARS courses
CEIBO	**	Inf.	– Inter-institutional coordination with Bimonthly meetings NGOs and NARS	– Consultations about coffee with NARS		– NARS trained extensionists

PROCADE	**	F, Inf.	– Participates in NARS planning exercises			– Information on PROCADE's staff resources to NARS
Member NGOs of PROCADE	*	Inf.		– Coordination for forage research		
GIA	***	Inf.	– Ex-GIA staff join NARS – Joint committee – Carried out diagnosis for NARS	– NARS contracted NGO for extension		– NGO trained NARS
AGRARIA	***	Inf.	– Joint preparation of research project	– Joint trial – NARS contracted NGO for extension		
(2) Public-sector cases						
CIAT	***	F	– Zonal meetings with NGOs – Joint diagnosis	– NGOs participate in adjustment trials – Joint workshops to determine recommendations		– Extension bulletins for NGOs – Information centre for NGOs – Field days for NGOs
INIAP-PIP	**	Inf.	– Joint diagnosis	– Exchange of technical information – NARS supplies seed to NGO – Joint adjustment and transfer activities – Joint planning of demonstration plots – NGO advises about adjustment trial location		

* little importance ** some importance *** important

[a] By adaptive trial type 1, we refer to those concerned with what we denominated "technology selection". By adaptive trial type 2, we refer to those concerned with what we denominated "technology adjustment" (see earlier discussion of phases of adaptive research).

Box 7.6 INIAP-PIP: achievements and limits of informal linkage mechanisms

The National Institute of Agricultural Research (INIAP) of Ecuador created a Programme of Production Research (PIP) with a systems focus. Each PIP Unit carries out on-farm trials for selection and adjustment of technology. The farmer participates in the preparation, installation and harvest of the trials. The aim is to transfer trial results to NARS and other extensionists. The programme is intended to generate feedback on farmers' needs in order to orient the work of INIAP's experimental stations.

Two of the ten PIP Units (Chimborazo and Cayambe) had worked with NGOs. The relationship was informal because no official agreements existed with the NGOs. In Cayambe, INIAP worked on separate occasions with CESA and CAAP, building on existing friendships between members of the different institutions that originated from university days. They discussed technologies with each other, PIP staff provided the NGOs with seeds of improved varieties, and technology-adjustment and -transfer work was carried out jointly. The PIP used the experience of NGO staff to locate their adjustment trials with farmers who worked with the NGO. The NGOs used the 'technological package' developed by INIAP in its 'production plots' to demonstrate the technology to farmers. NGO promoters assisted in transferring the technology to farmers. All these activities were supervised and managed by the NGOs and the PIP together. This coordination was made possible by the flexible working hours and work ethic of the PIP Units and the NGOs, in contrast to the difficult relationship between the PIP and the Ministry of Agriculture, reflecting the Ministry's inflexible working practices.

Active campesino participation in the work of the PIP Units was made possible both by the work of the NGO staff, who organized farmers for field activities, and the efforts of the PIP staff to reach the remote campesino communities where the NGOs worked.

However, there were also problems. One of the most important was the lack of any formal agreements, which made it impossible to demand the fulfilment of any commitments that were made. Furthermore, after about three years, collaboration between the PIP Units and the NGOs ceased when the staff involved, particularly those of the NGOs, changed.

This collaboration also showed the importance of individual initiatives within the NGOs, particularly in the case of CAAP. As we noted, at an institutional level CAAP had no policy to collaborate with INIAP; however, at a field level some of CAAP's staff who questioned this institutional decision took it upon themselves to engage in collaborative activities.

Source: Cardoso 1991; interviews.

Joint planning

Joint planning requires formal mechanisms so that decisions taken have the full support of the institutions involved. However, only two cases were found of formal planning mechanisms (CESA-Ecuador and CIAT). The participation of NGOs in planning meetings of the NARS, as in the case of CIAT, is

Box 7.7 Experiences of CIAT in the Mesothermic Valleys with joint problem-diagnosis

For a number of years CIAT has been employing the *sondeo* methodology developed by IICA in Guatemala (Hildebrand and Ruano 1982). In the *sondeo*, researchers analyse the different technologies that farmers are using, identify problems with them, and go on to consider the possibility of orienting future ATD work to address the problems identified. This type of *sondeo* uses a multi-disciplinary team made up of eight to ten people from different research departments.

In 1991 the CIAT organized a *sondeo* in the Mesothermic Valleys, which lie to the west of the department of Santa Cruz and are made up of a group of lower-altitude Andean Valleys. The extension staff of four Intermediate Users (IUs) participated in the *sondeo*, two NGOs and two public-sector agencies. Together with CIAT staff this made up a team of twenty-five people.

Field interviews with farmers were carried out by pairs which changed each day. A deliberate effort was made to ensure that IU and CIAT staff formed mixed pairs so that there could be a full exchange of views between the different institutions. Involving the IUs led to certain changes in the *sondeo* methodology. Previously, *sondeo* participants had discussed the day's findings in a single group and written preliminary findings jointly. Because of the number of participants this was not possible. Each pair summarized findings on a flipchart which was then presented to the full group.

After field interviews had finished, participants were assigned different themes in a provisional report. They presented these themes in a workshop held in the area. Managers of IUs and CIAT commodity researchers were invited to the workshop. While the physical output of the *sondeo* was a report, the less tangible but equally important result was to have brought together NARS and IUs during the days spent working together, visiting farmers and engaging in long discussions in the evenings about their practices and problems. This provided a strong base for future collaborative activities.

The contacts and conversations between technical staff based in the zone increased their knowledge of the work of other institutions and helped future coordination. In this sense the *sondeo* is a tool that is useful not only for collecting and analysing data but also for encouraging inter-institutional coordination and communication.

Source: Veléz and Thiele 1991.

potentially a useful mechanism for ensuring that research programmes respond to farmers' needs. It is, however, all too rare.

Two cases of informal joint planning were found in the study, although discussions in the workshop indicated that others do exist. In the case of AGRARIA, NGO staff prepared a research proposal with INIA investigators who collaborated on a personal and informal basis. At a higher level in the planning process, experiences accumulated by GIA in its project in the Diguillín region of Chile were passed to INDAP when the field leader of the project and other ex-GIA staff were contracted as part of INDAP's Regional

Management in 1990. Working together with an INDAP team, they have helped to reformulate elements of INDAP's strategy, making use of the experience of GIA and other NGOs (Sotomayor 1991).

Joint programming

Programming deals with the organization of specific activities over a relatively short period of time. In two of the case studies (El CEIBO and CIAT) programming was carried out in meetings in the area where NARS and NGO field staff worked. This kind of meeting served to build on complementarities between activities which had already been planned. It should be considered as a supplement to longer-term planning and not an alternative.

Joint evaluation

No examples were found in the case studies where NGOs and NARS had jointly carried out evaluation. In the absence of joint evaluation, feedback on problems with linkages between NGOs and NARS is weakened.

Professional activities

Professional activities cover those functions which are most closely associated with ATD, such as trials and demonstrations. They are analysed here according to the different functions distinguished in the ATS.

Applied research

Important interdependencies exist between research and transfer (see Box 7.4). Applied research requires links with transfer to ensure that the options being investigated are appropriate. If these links are not present, there is a high risk that the technologies developed will not be accepted by farmers.

In the case studies, no linkage mechanisms were found which related directly to applied research. Some linkage mechanisms which operate at the level of adaptive research affect applied research, but linkages with an actor in adaptive research guarantee neither coordination nor communication between that actor and others operating in other sub-systems of the ATS. Thus, for instance, in Ecuador, CESA established linkages with the PIP Units. However, because the PIPs did not have good contacts with the experimental stations, CESA's ability to influence applied research in INIAP via its relations with the PIPs was very limited (CESA 1991).

Adaptive research

Coordination of researchers with transfer agencies becomes even more important at the level of adaptive research. Feedback is needed on the behaviour of technologies under farmer conditions. Advice on trial design and assistance with interpreting results is also important. In addition, linkage mechanisms should pass information about the characteristics of new validated technologies from research to the actor who carries out transfer.

In adaptive research of type I, or what we have labelled technology selection, the technology is adapted to an agroecological zone. Only one linkage mechanism was found relating to this type of adaptive research: the joint management of an experimental station in Bolivia. Initially the mechanism did not work very well and had to be modified. Once modified, however, the mechanism improved the quality of feedback from the NGO to the NARS, as well as the flow of information from the NARS to the NGO about the technologies being selected (Box 7.8).

Box 7.8 CIPCA and CIAT jointly manage a local experimental station

While CIPCA has always seen type I adaptive research as important, it has come to believe that an institution which works directly with campesinos should not carry out research of this nature. CIPCA therefore only engages in such research when it has worked in areas where the NARS has no experimental station.

In 1976, when CIPCA began to work in the Cordillera, an area in the south of Santa Cruz bordering the Chaco, there was no research institution in the area. CIPCA therefore decided to establish its own experimental centre for research in fruit, pigs, cattle and grain crops at a place called Charagua. Subsequently, CIPCA reached agreement with CIAT for the joint management of research in the centre. The station became one of the seven Regional Research Centres (CRI) that CIAT used to adapt technology to regional agroecological conditions. CIAT provided a full-time researcher in the CRI and operating costs were shared with CIPCA. However, joint research covered only grain crops and livestock, because specialists in fruit production were not available in CIAT. Adjustment and transfer of technologies coming out of the CRI is carried out solely by CIPCA.

Initially, there were difficulties in the operation of the CRI because the researcher was responsible to one manager in the NGO and another in CIAT, and so felt pulled in different directions. Once this ambiguity was clarified, and the researcher was essentially given more autonomy, research progressed more smoothly. Once contacts between CIPCA and CIAT were formalized under the agreement that the CRI ought to be jointly managed, researchers from the main CIAT programmes travelled more frequently to the CRI in Charagua to support ongoing research. From CIPCA's point of view, this linkage thus helped draw CIAT resources and expertise into the area.

The close relationship between the CIAT researcher in the CRI and the CIPCA field team has been fruitful. The researcher informs the team about progress in research by means of monthly and annual reports. The CIPCA field team orient type I research by feeding back to the researcher the results of their technology adjustment work with campesinos. CIPCA field workers also make two formal visits a year to the CRI, organized by the CIAT researcher.

Source: Bebbington *et al.* 1992: 53; Garcia *et al.* 1991.

Type II adaptive research, or adjustment of technology, seeks to adapt the technology to the socioeconomic characteristics of a specific group of farmers. In this, the help of technology transfer agents is needed in the identification of representative locations for trials, and for trial design and evaluation. Involving transfer agents also helps them to learn about the technology being tested.

Most linkage mechanisms encountered related to adaptive research of type II, probably because coordination is easier when both actors are working in the field at an on-farm level where the benefits of collaboration are more readily apparent.

Frequently these linkage mechanisms are informal. Sometimes the NARS benefits from the local knowledge of NGOs to improve trials (as in the case of CAAP and INIAP). Sometimes, when the NGO carries out the trial, it informally seeks advice from the NARS about the technology being tested.

CESA-Bolivia developed special linkage mechanisms to support its adjustment trials. The mechanism was formal but not permanent, consisting of field visits by the NARS staff and study visits by paratechnicians trained by the NGOs to the NARS (see Box 7.9). This mechanism supported technology development following Path 3, with the NGO adjusting technology developed by the NARS.

Another example of a formal linkage mechanism for the adjustment of technology is the case of CIAT-DTT. Here it was the NARS which took the initiative and sought to involve NGOs (see Box 7.10).

Technology transfer

Three cases were found (CESA-Ecuador, IDEAS, INIAP-PIP) where there had been some loosely specified collaboration for technology transfer. Only in the Santa-Cruz area of Bolivia and Chile did well-developed mechanisms appear to exist.

In Bolivia the mechanism is voluntary. The NARS Department of Technology Transfer (DTT) has organized workshops and invites extensionists of different organizations, including NGOs, and investigators to determine cropping recommendations to be published in technical bulletins.

In Chile the mechanism is mandatory. INDAP contracts NGOs to provide technical assistance to farmers (Box 7.11). As we discussed in Chapter 6, NGOs have argued that this mechanism requires modification and have sought, with some success, to modify it.

Resource allocation procedures

It is possible to support the formation of linkages by formally assigning staff time, funds or other resources to linkage activities. Only one instance of this type of linkage mechanism was encountered. In Ecuador, with the prompting of a donor agency, INIAP lent an expert in fruit production to CESA on a part-

Box 7.9 CESA-Bolivia: non-permanent mechanisms for coordinating adaptive research (Path 3)

CESA wanted to implement an agroecological project in Yucumó, a new colonization zone. There were no institutions working in the area who could provide advice on agroecological methods and CESA lacked funds to contract experts in this field. This obliged CESA to establish contacts with different institutions involved in tropical agriculture in other parts of Bolivia.

Experts were invited for brief periods to advise the field team and to give short training courses. Experts came from the NARS (CIAT) and its bilateral support staff (British Tropical Agricultural Mission), other public-sector institutions and NGOs. Relations with the NARS were also very important in the supply of seeds and tree seedlings for use in agro-silvipastoral systems.

As well as visits by experts, groups of campesino paratechnicians from the project have made study visits to the research stations and experimental plots of CIAT in Santa Cruz and IBTA in the Chapare. This created an informal network of contacts, and enormously enriched the agroecological knowledge base being applied in the project.

By means of these temporary links, CESA managed to incorporate new technologies in its project, while it was being implemented. Since the technologies came from other regions and had not been tested in the zone, the project in effect became an experimental site.

In order to improve the coordination of these activities and the exchange of information generated, institutions in Yucumó created a local network. The network organized an agroecological symposium in which the external experts and other local institutions participated. The symposium served to consolidate local thinking about agricultural development strategies, and provided a base for subsequent project design in CESA.

Source: Kopp and Domingo 1991.

time basis (see Box 7.16 on CESA-INIAP). This mechanism contributed to the success of this particular collaborative programme. The donor's presence helped assure that resources were assigned to this mechanism, in contrast to other CESA-INIAP linkage activities which have, as a result of resource scarcity, had limited success.

In other instances, linkage mechanisms similarly failed because no resources had been allocated to support them. Linkage mechanisms have costs which must be anticipated and met if they are to work properly.

Communication devices

The case study on protected horticultural systems introduced by NGOs in Bolivia (see Ch. 5) concluded that a fundamental failing of NGOs is that they do not make their research results available to the NARs. Generally, information coming out of the NGOs is in pamphlet form and is so simplified for use by the NGOs' own project that it is not appropriate for researchers'

Box 7.10 CIAT-DTT: NGOs participate in adjustment trials (Path 2)

CIAT's Department of Technology Transfer (DTT) has carried out adjustment trials with the collaboration of NGO extensionists. The subject-matter specialists of the DTT organize these trials, which are called 'participatory tests'. They are tests, because although the technology has generated successful results in local research centres, it has not been proven under farmer management. They are participatory because the NGO extensionist and the campesino each play an active role in the process.

Participatory tests in agroforestry systems started at the end of 1990. They include forage alleys, windbreaks and living fenceposts. They are based upon research which CIAT began in 1986. By the end of 1991 a total of twenty-nine tests had been established in various communities occupied by settler farmers with different Intermediate Users (IUs), including seven NGOs and other institutions.

The design of the tests is simple, with one new technology being tested against the farmer's practice. For example, with the forage alleys, rows of the leguminous shrub *Leucaena leucocephala* are established with pasture in one half of the plot, and the other half is sown only with pasture. Cattle are grazed in both halves for the same length of time.

The collaboration of the extensionists of NGOs and other IUs is important because of their knowledge of the area and the community where the test will be established. However, the quality of support received from extensionists has varied considerably. In some cases planning, organization, sowing and other tasks were carried out by the IU. In other cases the extensionist only visited the test in the company of the subject-matter specialist.

Participatory tests are a useful instrument for getting feedback from NGO field staff and campesinos, whose questions have indicated several modifications to improve the systems, which are passed on to the researchers. For example, campesinos asked if forage alleys could be adapted for sheep by pruning the *Leucaena* shorter – the test design was subsequently modified to take this into account.

Carrying out participatory tests helps farmers, NGO extensionists and CIAT's subject-matter specialists to understand the reasons behind each other's behaviour. This establishes a strong base for subsequent transfer activities.

Source: Veléz and Thiele 1991.

needs (Kohl 1991). A converse problem exists in the NARS. Frequently they only publish research results and not more practical information suitable for the transfer activities of NGOs. It is rare for either NGOs or NARS to use communication devices explicitly targeted at the needs of the other.

The lack of communication devices as explicit linkage mechanisms can be seen in Table 7.3. The main exceptions were related to the work of CIAT-Bolivia, where a new model had been introduced which sought to develop linkages with the NGOs. However, there are certainly many more instances in which communication devices transmit information from NARS to NGOs without any explicit targeting. NARS manuals are often found on the shelves of NGO staff. The converse is far less frequent.

Box 7.11 Chile: INDAP contracts NGOs to carry out extension

In its Programme for Technology Transfer, the Chilean government institution for technology transfer, INDAP, provides extension to farmers through a number of 'private consulting companies'. The activities of the companies are financed by a government subsidy for each subsistence farmer who participates. INDAP plans the allocation of resources and monitors the extension programme.

From 1990 INDAP has allowed NGOs to participate as consulting companies. The NGOs have criticized various aspects of the Programme. They say that it is designed for individual farmers, covers only technical assistance and is not related to credit programmes, and is both top-down and short term.

Two of Chile's largest NGOs, AGRARIA and GIA, have become involved in this programme. Their motivations for doing so, and scale of involvement, have however differed somewhat.

AGRARIA has become massively involved in the programme and in 1991 gave technical assistance to 2,200 farmers with INDAP financing. One of the motivations for doing this has been to have a significant influence over the nature of work done in INDAP programmes, by executing them itself. Another has been to try and use the contracts as one basis on which to negotiate changes in the programme's design. However, the strategy of massive involvement has introduced certain tensions in AGRARIA, which has grown very rapidly to absorb these contracts and has had difficulty controlling the quality of its own work. It is also plagued by the nagging doubt that it is becoming an executing agency for government.

GIA also decided to enter the Programme, but with a different emphasis. GIA's involvement is on a far smaller scale, and is mainly aimed at studying the dynamics of INDAP contracts. One of GIA's technical teams has become a consulting company with the objective of studying the Programme's operation and looking at its impact on campesino farmers. It plans to pass results on to INDAP so that the Programme can be improved.

In 1991, GIA took charge of three contracts in the Programme. Involvement in implementation has helped it to study a gradual transformation in the Programme. For example, it has made an attempt to carry out extension with groups and not individuals, a change which has since been recommended by INDAP management.

As a result of participating in this programme, the technical teams of GIA are in constant contact with INDAP. Implementing these contracts has been useful for GIA to understand the Programme and to establish a dialogue with INDAP.

It would seem that this smaller-scale involvement has been easier for GIA to absorb than has been the case in AGRARIA. Conceiving a contracted relationship to the state as a research enterprise, rather than a basis of institutional expansion, is far more consistent with the history and identity of the NGO.

Sources: Aguirre and Namdar-Irani 1991; Sotomayor 1991; AGRARIA 1991.

In the CIAT-DTT the principal communication devices which target NGOs are:

- bulletins and other technical material for extensionists;

Table 7.3 Structural linkage mechanisms

Institution	*Type of mechanism*	*Mechanism*
INIA and NGO, Chile	Permanent committee	Area commission
ICA, Colombia	Permanent committee	Consultative council
CIAT, Bolivia	Coordination unit	Department of Technology Transfer
PROCADE	Coordination unit	Central office
IBTA-PROCADE	Representation of one actor in the governing body of another	Executive council

- information centre open to extensionists;
- training courses for extensionists;
- field days for extensionists.

In the case of GIA, the communication device is training courses targeted at the NARS, a device used also by other NGOs. Since the restoration of democracy in 1990 GIA has trained staff of INDAP and INIA in areas such as methodologies for problem diagnosis, campesino production systems and how to work with campesino organizations.

EXPERIENCES WITH STRUCTURAL LINKAGE MECHANISMS

The formation of structural linkage mechanisms and the creation of NGO networks has been associated with recent radical change in some of the NARS. Given that these are recent, it is not surprising that relatively few structural linkage mechanisms were found and all of these were in a process of consolidation. Table 7.3 shows examples of the three types of structural linkage mechanism which were found.

The permanent committee

The permanent committee is made up of representatives of both actors. Theoretically, these committees could have the power to take mandatory decisions, or they could be voluntary in the sense that they only offered advice. The only example of a permanent committee which really functions as a linkage mechanism is the CATT in Chile (Box 7.12). Here the NGOs participate in committees at regional and area levels. These are mandated mechanisms because the committee participates in the preparation of the Action Plan which NARS staff are then obliged to follow.

Box 7.12 GIA and the centres for adjustment and transfer of technology

In 1990 INIA began to establish fifty-five centres for adjustment and transfer of technology (CATTs). Each CATT is located in an agroecological zone.

GIA signed an agreement to participate in the regional commission (committee) of the 8th Region and in three area commissions (committees) and has sought to take advantages of these new opportunities for NGOs to influence INIA's agenda. It has contributed to the design of the CATT programme at three levels: at a *national level*, it contributed to the design of the main methodological elements of the CATT programme; at a *regional level*, it worked in the identification of broad agroecological areas in the 8th Region of Chile, and on the spatial distribution of the CATTs in these areas; and at a *local level*, it concentrated on the practical aspects of establishing three CATTs, which included the selection of plots and the identification of research topics and of specific methods of operation.

Source: Sotomayor 1991.

In Colombia, the NARS ICA has recently created consultative councils (Box 7.13) which could potentially function as linkage mechanisms with NGOs. These councils operate at the level of the Regional Centres for Extension, Training and Diffusion of Technology (CRECED). The CRECEDs are decentralized centres for the adjustment and transfer of technology, similar in some ways to the CATTs in Chile. The councils involve Intermediate Users but NGOs are very weakly represented and have little contact with the CRECED. They represent a linkage mechanism whose potential is yet to be exploited.

The coordination unit

Coordination units have no formal authority over the institutions whose actions they coordinate. By definition they are voluntary mechanisms. They may form part of one or other of the institutions involved in the link or they may be independent (Merrill-Sands and Kaimowitz 1990: 41). Coordination units, or coordination positions, have been implemented in the public sector, for example in Nigeria (Ekpere and Idowu 1989) and Zambia (Kean and Singogo 1990). The clearest example of a coordination unit found in the case studies is the Department of Technology Transfer (DTT) of CIAT (Box 7.14). This was formed with the express objective of establishing linkages between CIAT research and the extension being carried out by NGOs and other IUs.

The localized nature of NGO activities has led to their grouping together in networks or forming coordinating bodies (chapter 6). When these explicitly seek to develop relations with the public sector they can also be classified as coordination units, even though they have other functions. In Bolivia,

Box 7.13 ICA: consultative councils – a potential linkage mechanism with NGOs

In Colombia, ICA has formed regional centres for extension, training and development of technology (CRECEDs). ICA is now trying to decentralize its activities, increase farmer participation and develop linkages with other institutions and within ICA itself. The CRECED has four units with specific activities:

- adjustment and validation;
- agricultural services;
- diffusion;
- evaluation and monitoring.

The adjustment and validation unit carries out on-farm research. The diffusion unit combines the results of technical research with farm management concerns. It then transfers results to the Intermediate Users who include private professional advisers, public-sector projects, campesino and large farmer leaders, and occasionally NGOs.

Each CRECED has a technical committee made up of unit coordinators and the CRECED director. Decisions taken by the technical committee have to be approved by a consultative council. In the Meseta Popayán CRECED in southern Colombia, the council is made up of representatives of private professional farm advisers, professional associations, public-sector institutions and farmer groups, as well as the CRECED director and a coordinator whose job it is to develop linkages with the Palmira Experimental Station. The CRECED and the consultative councils could become important local linkage mechanisms with NGOs without further radical changes in the structure of ICA.

Source: Estrada 1991.

PROCADE has acquired some of the functions of a coordination unit, with some autonomy from the NGOs which it links.

The experience of PROCADE shows that establishing a structural linkage mechanism is not in itself sufficient to establish functioning links. It is necessary to complement it with specific operational linkage mechanisms (Box 7.15).

THE LIMITS OF INTER-INSTITUTIONAL COORDINATION

The types of linkage mechanism which might most successfully be used in a given case depend upon the characteristics of the ATS and contextual factors. In Chile and Santa Cruz, Bolivia, structural linkage mechanisms have been introduced and operational linkage mechanisms developed. However, in these cases contextual factors have been more favourable to NGO–NARS linkages.

In other places it has been more difficult to introduce linkage mechanisms with wide coverage. This is illustrated by the experience of CESA and INIAP in Ecuador (Box 7.16). After recognizing the differences of interest between

Box 7.14 The CIAT Department of Technology Transfer: complementing a coordination unit with operative linkage mechanisms

In 1985 CIAT closed its extension department which had worked directly with farmers (see Box 6.13). It now depends for technology transfer upon the links established by its coordination unit, the Department of Technology Transfer (DTT), with organizations it terms Intermediate Users (IUs). These have extensionists who work with farmers and act as intermediaries between CIAT, campesinos and commercial farmers. In Santa Cruz, NGOs are one of the most important types of IU.

The DTT has subject-matter specialists and zonal specialists. Their work is supported by a communications section. The subject-matter specialists are in constant contact with their corresponding CIAT researcher and collaborate on some research work. They package research information for delivery to IUs and should transmit feedback on farmer needs to the researcher.

Frequently, technologies developed in the experimental centres are still not ready for transfer. One of the responsibilities of the subject-matter specialists is to carry out on-farm adjustment trials. Another of their functions is to inform and train extensionists to ensure that they pass on the appropriate technological message to farmers. They prepare technical bulletins for extensionists using research results. They advise extensionists on some aspects of their work such as the establishment of demonstration plots and how to give talks to farmers. Finally, they play an important role in feedback, communicating technology needs of IUs and farmers to the investigator.

Each zonal specialist is designated a geographical area within which the specialist is responsible for developing and coordinating technology transfer activities by the different bodies which work there. Zonal specialists should look at transfer with a systems focus, integrating the perspectives of the subject-matter specialists.

Zonal specialists organize DTT support for IU extensionists in their zone. They are responsible for coordinating with IUs in *sondeos* and other diagnostic activities carried out in the zone (see Box 7.7). They should ensure that the extensionists receive the technical information they need from CIAT and the support of the appropriate subject-matter specialists. It is their responsibility to organize zonal technical meetings with CIAT staff and extensionists from many IUs. These serve to:

- share the knowledge of CIAT and the IUs about the zone;
- identify training needs and train extensionists;
- coordinate plans and activities;
- provide feedback to research.

Goals for this model of operating have been defined but the specific mechanisms and means to realize the goals are continually adapted and developed in the course of activities which link the different institutions involved. It is necessary to develop the coordination mechanisms in more detail.

Source: Veléz and Thiele 1991.

the NARS and the NGO, CESA has concluded that only very specific linkage mechanisms should be implemented in situations where there was a clear coincidence of interest.

Box 7.15 PROCADE: a coordination unit on its own does not guarantee functioning links

The Alternative Campesino Development Programme (PROCADE) is a programme for inter-institutional coordination made up of a group of NGOs which work in five Andean Departments of Bolivia. PROCADE coordinates its member NGOs, and acts as an external representative with the public sector and bilateral agencies. It also has come to constitute a coordination unit in negotiations and liaisons with IBTA, although this was not the reason behind its initial creation.

PROCADE has a central office for National Coordination made up of a coordinator, an agronomist, an educator, an economist and administrator, an accountant and a secretary.

At the level of the central office, PROCADE has good relations with IBTA's National Programme leaders and researchers (in potatoes, quinoa, legumes, maize, wheat, livestock and forages). The Central Coordination of PROCADE represents the NGOs in the managing council of IBTA and has participated in high-level discussions concerning the formulation of a new research and extension strategy for pre-extension which IBTA is planning to implement.

However, coordination between the technical teams of the NGOs which make up PROCADE and IBTA is very limited. This is due to institutional jealousy on both sides and the lack of permanent and flexible mechanisms for inter-institutional coordination (there are no IBTA offices in areas where the NGOs work). This shows that the formation of a coordination unit does not resolve all linkage problems. It will be necessary to develop more specific operational mechanisms to coordinate different functions of generation and transfer.

Source: Gonzalez 1991.

CHAPTER SUMMARY

Principal findings

We began this chapter describing different paths for ATD. These represent different ways in which ATD functions can be divided between NARS and NGOs. In some instances (those denominated as Path 2), the role of the NGO was limited to transferring technologies generated by the public sector. However, Path 3 was more frequently encountered. Here the NGO carried out adaptive research and transfer activities and the public-sector applied research. This path has predominated because NGOs work in marginal areas, and NARS technologies are typically developed with resource rich farmers in mind, so that some adaptation is required. Path 8, where the NGO carried out applied and adaptive research as well as transfer, was also important. This path occurs where the NARS is ineffective in the zone or the technologies it offers are not those the NGO wants, as in the case of some agroecological NGOs.

The division of functions between NARS and NGOs within the ATS which occurs in most paths shows that there are interdependencies between the two types of actors. The experiences analysed in this chapter demonstrate that

Box 7.16 CESA and INIAP Ecuador: the limits of inter-institutional coordination

In Ecuador, CESA and INIAP established an institutional relationship in 1976. CESA wanted to gain access to technological resources and specialized scientists and direct them towards campesinos who had to date been marginalized from government support.

The two institutions worked both with formal and informal mechanisms (see Box 6.6). Formal mechanisms have been supported by specific agreements of different types:

- 1976: INIAP trained and advised CESA staff on the establishment of trials and demonstrations.
- 1979: the two organizations coordinated the establishment of demonstration-trials and experiments on campesinos' farms.
- 1981: joint execution of a soil and water conservation programme.
- 1982: cooperation in investigation, demonstration and extension of fruit technologies in the Chingazo-Pungal project in Chimborazo.

As a result of these experiences, CESA requested the establishment of a general agreement on technical cooperation, which was formalized in 1986.

The most successful of these mechanisms was cooperation in the fruit project in Chingazo-Pungal. In part, success was due to the influence of COTESU, the Swiss bilateral funding agency. COTESU was financing an INIAP project to strengthen research in fruit. At the same time COTESU supported a project with campesinos in Chingazo-Pungal, where fruit was a potentially important crop. COTESU made use of its relationship with INIAP to encourage greater contact between the two institutions.

This three-way relationship created a space within which CESA could modify the high-cost technology packet being proposed by INIAP, and incorporate some innovations discussed with campesinos. INIAP provided CESA's project with a specialist in fruit production on a part-time basis. This was a significant help to the programme, although there were operational difficulties due to INIAP's lack of funds. In the case of this project, inter-institutional collaboration allowed farmers to incorporate and adapt the fruit production technology and led to substantial economic improvements.

In contrast with the success of this specific mechanism, the general agreement on technical cooperation has not had satisfactory results. According to interviews with managers in INIAP this has been because of:

- lack of specificity as to what the relationship should involve;
- imprecise objectives;
- inadequate designation of responsibilities and budget;
- lack of mechanisms for evaluation, control and monitoring.

A 1988 evaluation of CESA's ATD work concluded that the general agreement did not work well because of problems at INIAP, such as:

- rapid staff turnover;
- the export crop orientation of INIAP's research;
- a lack of willingness on both sides to analyse research results jointly and support research with peasant farmers.

After this experience, CESA has concluded that joint planning with INIAP is difficult to achieve: 'However, it is possible to agree on the planning and execution of some activities when they have a specific and concrete objective, under certain conditions and certain circumstances'.

Source: CESA 1991.

these interdependencies are not simply *potential* complementarities but reflect *real* complementarities that have been realized in prior experiences. In order to take further advantage of these real complementarities and ensure that the paths function efficiently, coordination between actors is required. This implies a need for more explicit linkage mechanisms.

In the case studies we found that, frequently, the institutions involved did not recognize the existence of the linkage mechanisms or paid little attention to their organization. As a result many are informal, poorly structured and needlessly inefficient.

The linkage mechanisms identified were defined according to two main categories:

- Operational mechanisms, which coordinate specific ATD functions.
- Structural mechanisms, which create linkages at a general planning level.

The most common operational mechanism was 'Collaborative Professional Activities' which covers those functions most closely associated with the ATD such as trials and demonstrations. Linkage mechanisms for trials mostly related to type II adaptive research, for the adjustment of technology on-farm to the characteristics of a specific group of farmers. This was probably due to the fact that when both actors are working in the field coordination becomes easier and the benefits more apparent.

Of the other operational mechanisms, 'Planning and Review Processes' (which included joint problem-diagnosis, joint planning and joint programming), were found in half the case studies, but no instances were found of joint evaluation. This is a significant weakness as such joint evaluations would provide particularly effective ways of improving inter-institutional coordination, communication and understanding. Similarly, 'communication devices' were found in only a third of the cases. Both NGOs and NARS rarely use communication devices which explicitly target the needs of the other, with the result that communication between them is usually poor. Finally, 'resource allocation procedures' was only encountered once as a linkage mechanism. The failure of institutions to allocate resources for the operation of linkage mechanisms limited their effectiveness in a number of instances.

Operational mechanisms can be either informal or formal. Informal mechanisms were found to be useful. They cut through bureaucratic procedures and exploit personal contacts. Probably the most important informal mechanism is the *ad hoc* consultation. Other mechanisms which require more sustained coordination, such as joint trials, can also be carried out informally. However, if there is a change in staff, informal mechanisms may break down. Furthermore, informal mechanisms are unable to resolve more profound problems in relations between actors in the ATS. For these reasons there may be advantages to formalizing linkages: the risk is that in doing so much of the flexibility of informal relations is lost.

Three different types of structural linkage mechanism were found:

- Coordination units.
- Permanent committees.
- Representation in managing bodies.

At the time of the research, structural linkage mechanisms had only been introduced and initiated in Bolivia and Chile following recent radical changes in the institutional structure of their ATS. It is not yet possible to make a definitive evaluation. However, several points have become clear. First, structural mechanisms can provide the basis for better coordination between the two actors. Second, establishing a structural linkage mechanism is not in itself sufficient for links to function, they must be complemented with operational linkages. Third, establishing structural linkages helps the formation of operational ones. Finally, we observed that linkage mechanisms with a wide and general jurisdiction do not always work. Where there are marked differences of interest between the NARS and the NGOs, only very specific linkage mechanisms are effective, where the needs of both clearly coincide.

Overall, it is clear that the establishment of structural mechanisms requires much more time, effort and willingness to negotiate on the part of the institutions involved than does the establishment of operational mechanisms. Yet ultimately it seems that a more broad-based participation of NGOs and farmers' organizations in resource allocation and policy decisions in NARS (i.e. the democratization of those NARS) will require structural mechanisms.

Implications of the findings

The overall implications of the patterns identified in this chapter seem to be as follows:

- Past experience suggests that there is considerable potential for linkage mechanisms that will improve the functional linkages within the ATS and so improve the quality of ATD for campesinos. The mechanisms could, however, be greatly improved if they were developed and managed more actively.
- Despite NGO criticisms of the emphasis of NARS and donors on improving functional links between NGOs and the government, NGOs have in practice been willing to engage in such relations in the past without saying so explicitly.
- The albeit limited evidence suggests that structural linkages can only be introduced when special conditions are met in the ATS and the broader political arena (as in contemporary Chile). Elsewhere it is going to be far more difficult to democratize decision-making in ATD and increase the involvement of institutions of civil society in defining and monitoring public policy.

Critical considerations in establishing linkage mechanisms

The evidence presented in this chapter suggests that in practice NGOs have recognized a role for linkage mechanisms, even if they have not accorded them the importance or resources they merited. These sentiments were endorsed in the workshop in Santa Cruz.

The chapter and the workshop also suggest several reasons why far more attention should be paid to managing these mechanisms than has been the case in the past. There are several reasons why this should be so:

- A range of linkage mechanisms is needed to perform different functions. Which mechanism is most appropriate to the task at hand should therefore be carefully considered.
- However, these linkage mechanisms do not function automatically. Once established they have to be adjusted to fit the needs and capabilities of the different actors. This is a difficult task which requires skilled intervention on the part of managers.
- Moreover, linkage mechanisms are not cost-free: they require resources. Provision in terms of staff time and money should be planned. They therefore involve trade-offs, as resources spent on linkages will be taken away from other activities. Such trade-offs need to be carefully weighed by managers.

A second key observation is that with few exceptions the number and range of linkage mechanisms functioning at present is insufficient to allow markedly better coordination between the two actors. If complementarities are to be fully exploited, the number and range of linkages needs to be very substantially increased. In particular, the following merit greater attention, as they could improve the functioning of other linkages and have a multiplier effect in other parts of the ATS:

- Mechanisms for joint planning and for the allocation of resources.
- Communication devices which explicitly target the other actor. These are poorly developed, but are of critical importance in allowing information about technologies to flow within the ATS. Managers of NGOs and NARS should make a special effort to ensure that communication devices which supply the information required by the other actor are produced in a form which they can easily use.
- Structural linkage mechanisms can improve the relationship between NARS and NGOs, and facilitate the establishment of a set of operational linkage mechanisms. However, their establishment is not always possible, and is not sufficient in itself to create functioning links. They must be modified as circumstances dictate, and complemented with operational linkage mechanisms.

One final observation should be borne in mind by all managers. Linkage mechanisms which are established by the central office may have little effect on field-level staff. It is often best to begin locally when establishing or developing linkages. These can then respond to a more clearly perceived need, for it is in the field and on-farm that gains from coordination are most apparent.

In Chapter 9 we return to some of the wider implications of these patterns for the future of NGO–state relations in ATD. Before that, however, we turn to a chapter which considers the main themes in the experiences of Central American NGOs in ATD. This material, albeit in highly summarized form, helps provide both comparisons and contrasts with that concerning South America.

8

NGOs, THE STATE AND AGRICULTURE IN CENTRAL AMERICA

David Kaimowitz
Inter-American Institute for Cooperation in Agriculture

As in other regions of Latin America, the role of NGOs in agriculture in Central America has expanded markedly during the past fifteen years. Over 200 NGOs in the region now have agricultural projects. Together, they provide services to as many farmers as do government agencies, perhaps more.

NGOs have flourished in Central America for reasons similar to other regions: declining confidence in and capacity of public-sector agencies; emerging social movements that seek to institutionalize their efforts; growing concern within Catholic and Protestant churches for social issues; increasing donor and professional interest in participatory methodologies and appropriate technologies. Despite these similarities, however, the military, political and social conflicts which pervaded Central America in recent years give NGOs' development in Central America certain specific characteristics, and greatly influence relations between NGOs and the state.

This chapter provides an overview of NGOs' involvement in agriculture in Central America and their relations with the state.[1] It argues that major opportunities are emerging for coordination between NGOs and public agencies, both with regard to agricultural technology, and concerning broader social issues, which have been difficult to discuss constructively in these highly polarized societies. These opportunities could remain unfulfilled promises, however, if financial support for the public sector and NGOs continues to decline.

The term NGO, as used in this chapter, refers to private non-profit institutions, both national and international, which provide direct services to farmers, but are not themselves farmer organizations.

The chapter first discusses the main technological and social changes in Central American agriculture in the 1970s and 1980s, then examines how the government agencies and the NGOs evolved in that context. This lays the background for discussing the NGOs' current approach to agricultural development, their links with public agencies and international agricultural research centres, and the prospects for the future.

THE PATTERNS OF TECHNICAL AND SOCIAL CHANGE IN AGRICULTURE

Central American agriculture experienced rapid growth during the 1970s (about 3 per cent per year), stimulated by high international commodity prices, easy access to credit and the incorporation of large new areas into production (Bulmer-Thomas 1987). Growth was concentrated in bananas, cattle, coffee, cotton, and sugar produced for export (Williams 1986). Foodstuff production grew more slowly.

The expansion of export agriculture displaced large numbers of small foodstuff producers and promoted the concentration of landholdings (Brockett 1988). This, combined with demographic pressure from rural population growth, reduced many farmers' access to land. Most small farmers found it increasingly necessary to seek, low paying, off-farm employment in the export harvests or non-agricultural activities. Thus, despite economic growth, widespread rural poverty persisted.

A small but significant percentage of small farmers did manage to prosper and capitalize their farm (Baumeister 1987). These farmers were able to take advantage of proximity to urban markets for vegetables or dairy products, grew export crops such as coffee, had access to irrigation or high quality lands, or benefited from land rents obtained by bringing new lands into production on the agricultural frontier.

Small farmer heterogeneity was also reflected in the patterns of technological change. Average corn and bean yields rose slowly between 1970 and 1980 (Coutu and Gross 1990). Nevertheless, these averages disguise large disparities between the performance of small farmers on marginal lands, with little access to credit, whose yields stagnated, and others whose privileged access to high-quality land and credit allowed them to adopt improved varieties and greater agrochemical use and to increase their yields. The averages also mask a tendency among small farmers to use more fertilizers and pesticides to compensate for declines in productivity produced by soil degradation and growing pest problems.

With the military conflicts in Nicaragua, El Salvador and Guatemala, and the sharp decline in international commodity prices in the 1980s, agricultural growth slowed. Food production was less adversely affected than traditional exports, but even there yield and area increases were modest.

One of the few commodity groups which grew rapidly in the 1980s were high-value non-traditional exports of fruits, vegetables, flowers and ornamental plants. Most of these exports came from Guatemala and Costa Rica, and about one-third were produced by small farmers (Kaimowitz 1992).

Nicaragua and El Salvador carried out major agrarian reforms during this period. Large areas of land were expropriated and turned over to landless families, to be farmed collectively or individually. Similar reforms took place in Honduras in the 1970s. By 1987, approximately 100,000 of the region's 2 million farm families were in agricultural production cooperatives in the agrarian reform sector (Devé 1987).

The overvaluation of most Central American currencies during the first half of the 1980s reduced the price of imported agricultural inputs in real terms. In Nicaragua in particular, input prices dropped to very low levels and input use rose accordingly (Sthaler-Sholk and Spoor 1990). This trend was reversed at the end of the decade, when currency devaluations and reductions in agricultural credit limited farmers' abilities to purchase improved varieties and agrochemicals.

TRENDS IN THE PUBLIC-SECTOR INSTITUTIONS AND THE NGOs

Public research and technology transfer

The public-sector institutions involved in the Agricultural Technology System (ATS) in Central America include: Ministries of Agriculture, semi-autonomous research institutes, agrarian reform institutes, rural credit programmes, rural development projects, vocational training institutions, irrigation agencies, agricultural schools and universities and others. Resources for research and technology transfer for food products have traditionally been concentrated in the research institutes and Ministries of Agriculture – except for Honduras, where regional rural development projects have been important.

Most research in these institutions has been devoted to testing imported crop varieties under local conditions (Consultores SETA 1988). Transfer activities have concentrated on promoting the use of improved varieties and agrochemicals, often through subsidized credit or donations.

From 1950 until 1980, the human, financial and physical resources assigned to research and extension activities in the public sector rose steadily (Lindarte 1990). The effect of these efforts on small farmers, however, is uncertain. The creation and promotion of hybrid maize varieties in El Salvador in the 1960s had a major impact on small-farmer maize yields (Rodriguez, Navarro and Sanchez 1984). Similar results were achieved with open pollinated maize varieties in parts of Guatemala (Echeverría 1990). There was less success with beans and sorghum produced for human consumption. The public sector aided the growth in agrochemical use, but its exact contribution is difficult to determine.

Whatever their limitations in the 1970s, for most public research and technology-transfer institutions things became significantly worse in the 1980s. Widespread violence and declining real salaries and operating funds led many of Central America's best professionals to leave government or even the country. In El Salvador and Nicaragua, technology-transfer activities were largely abandoned in favour of efforts to organize producer cooperatives in the agrarian reform sector. Frequent organizational restructuring in Costa Rica, El

Salvador and Nicaragua paralysed institutional activities on many occasions (Palmieri 1990). Large loans for research and extension in Costa Rica and Guatemala from the Inter-American Development Bank (IDB) and the International Fund for Agricultural Development (IFAD) in the mid-1980s temporarily alleviated the public sector's decline. But the situation once again begin to deteriorate upon the projects' completion.

The crisis in the public institutions accelerated after 1988, due to reductions in US foreign assistance, termination of the IDB loans, and policy decisions designed to reduce public spending. In Guatemala, spending on public research and extension declined by 48 per cent and 60 per cent respectively between 1989 and 1991. The number of researchers in Honduras fell from 127 in 1990 to 78 in 1992, and extension agents were reduced from 561 to 241. Similar reductions took place in Nicaragua (Grupo Regional de Fortalecimiento Institucional 1992). The region's agrarian reform institutes, rural credit agencies and rural development projects also greatly reduced their technical assistance.

For the most part, there has been no clear strategy as to how to adjust the public institutions' mandates, structures, and methodologies to the new circumstances. Instead, these institutions have tended simply to drift along, with ever fewer resources and credibility.

Non-governmental organizations

As for the NGOs, only a few were involved in agriculture prior to 1970. Most of those were international NGOs, that were charity-oriented or based in the Catholic Church. Among the most prominent were the American Cooperative for Remittances to the Exterior (CARE), CARITAS, Catholic Relief Services (CRS) and Popular Cultural Action of Honduras (ACPH).

Natural disaster in Nicaragua (1972 earthquake), Honduras (Hurricane Fifi in 1975) and Guatemala (1976 earthquake) brought new international NGOs to those countries. Many stayed on and made the transition from disaster relief to rural development. Protestant churches of various denominations and political persuasions became increasingly active. Along with Catholic agencies, these churches provided important support to the growing peasants' movements.

At first the NGOs devoted little attention to agricultural technology. Their agricultural projects concentrated on donations and credit for agricultural inputs, organizing cooperatives and attempts to promote community gardens and small animal projects. Few of these initiatives had a sustainable impact.

Two NGOs which focused more on technology were the Mesoamerican Center for the Study of Appropriate Technology (CEMAT) and World Neighbors in Guatemala. CEMAT has worked since 1975 studying and promoting technologies such as improved wood stoves, biodigestors, solar driers and medicinal plants. World Neighbors has one of the oldest and most successful efforts in Central America to promote soil and water conservation techniques.

World Neighbors was also among the first NGOs to participate in a joint project with a public research agency. In 1974 it signed a cooperative agreement with Guatemala's Agricultural Science and Technology Institute (ICTA) to work together in San Martin Jilotepeque. Three community leaders who worked with World Neighbors were put on ICTA's payroll as part-time staff and participated in ICTA's on-farm trials for several years (Ortiz *et al.* 1991).

After the triumph of the Sandinista Revolution in Nicaragua in 1979, NGO efforts in Central America, both national and international, expanded greatly. Of seventy-one NGOs in Honduras identified in 1992 as having or having had agricultural projects, 64 per cent were formed after 1980 (Kaimowitz *et al.* 1992). In El Salvador the percentage was higher (Rodriguez 1991).

Many new NGOs were formed with ideological motivations and actively supported either the Left or the Right in the region's military and political conflicts. Activists on the Left, including many donor agencies from western Europe and Canada, saw support for the NGOs as an opportunity to create institutional structures independent from and critical of national governments (except in Nicaragua). The Right, often associated with USAID, saw the NGOs as private mechanisms for carrying out development activities more efficiently than the public sector. They implicitly assumed that these development activities would facilitate the counter-insurgency efforts of national governments (Barry and Preusch 1988).

The NGOs' political polarization was reflected in the national bodies created to promote coordination between them (Reuben 1991). USAID sponsored coordinating bodies in Costa Rica, Guatemala and Honduras that brought together and funded NGOs sympathetic to its aims. Separate, alternative left-wing coordinating institutions were set up in Costa Rica, El Salvador, Guatemala and Nicaragua. Thus, the coordination between NGOs was focused more on mobilizing political support and obtaining funds than on cooperating to find ways to provide services more efficiently or on exchanging technical information.

A massive influx of development assistance from foreign governments and NGOs, stimulated by the regional crisis, financed the NGOs' expansion. Donor support for NGOs in Central America went from almost nothing in the mid-1970s to an estimated 200 million dollars a year in 1987 (Levitt and Picado 1989). Many new NGOs were created or came to the region whose primary interest was gaining access to these funds.

The NGOs grew more slowly in Nicaragua and Costa Rica than in the other countries. Although some western European NGOs and solidarity organizations did conduct development projects of their own, most donor agencies sympathetic with the Sandinista Revolution preferred to finance government projects or the popular organizations affiliated to the Sandinista Front, rather than NGOs (Barraclough *et al.* 1988). The lack of US assistance to Nicaragua inhibited the expansion of anti-Sandinista NGOs. The relative lack of NGOs in Costa Rica is largely due to the comparative strength of the country's public sector, its higher standard of living and its less direct role in the military conflicts.

NGO–public-sector coordination

Coordination between NGOs and public institutions was minimal and largely informal. Many national NGOs perceived the government as a political opponent. Others were sympathetic to government aims, but viewed the public institutions as inefficient and unreliable partners. The public research and technology-transfer institutions were frequently almost unaware that the NGOs existed. Often they perceived those NGOs of which they were aware as overly paternalistic or radical and were jealous of the international NGOs' privileged access to resources.

During the early 1980s the military governments in Guatemala repressed community organizations and cooperatives promoted by the NGOs, prompting a number of international NGOs to leave the country. Despite continued violence in certain areas, the situation improved after the return to civilian government in 1986, and NGO activities once again began to expand (Matheu 1991). NGOs associated with the social movements in El Salvador also suffered government repression throughout the 1980s (Box 8.1).

In recent years, the most fundamental change affecting the NGOs has been the progress in overcoming the military conflicts in Nicaragua and El Salvador through negotiation and dialogue. The trend towards peace in the region has opened new opportunities for collaboration between the NGOs and the public sector. Ideological polarization has not vanished, but it has become less confrontational.

After the electoral defeat of the Sandinista Front in Nicaragua in 1990, many Sandinista professionals left government and joined NGOs. Between March and July 1990, the number of NGOs with agricultural projects rose from around fifteen to over forty. The government of Violeta Barrios de Chamorro which replaced the Sandinistas has maintained a policy of national reconciliation, which has permitted many Sandinistas to continue to occupy positions in the Ministry of Agriculture and Livestock (MAG) and the Natural Resources Institute (IRENA) and has made it possible for public institutions to work together with Sandinista-inspired NGOs.

The signing of the peace accords in El Salvador in early 1992 has also greatly facilitated collaboration between NGOs and the public sector. Soon after the accords were signed, the former Salvadorean guerrillas formed several new NGOs. Other NGOs in which they had been involved clandestinely have become more active. The National Reconstruction Plan which grew out of the negotiations between the Salvadorean government, the Farabundo Marti

Box 8.1 The changing relations between the Salvadorean government and leftist NGOs

Before the peace accords were signed in El Salvador there were five or ten NGOs in the country sympathetic to the FMLN, many of them associated with Catholic or Protestant churches. These NGOs had almost no relations with the Salvadorean government at the national level, were frequently victims of government repression and were extremely suspicious of government initiatives. Some of them operated mostly in areas controlled by the FMLN, where there was no non-military government presence.

With the signing of the peace accords things have begun to change. The FMLN has created its own unified NGO (the 16 January Foundation), and four of the five parties in the FMLN have also formed their own separate NGOs. Despite continuing strong opposition from certain groups within the Salvadorean government, it has largely accepted the idea that these NGOs will play an active role in providing a variety of social services in the 147 municipalities included in the national reconstruction plan (PRN) where the FMLN has an important presence. Nineteen NGOs and popular organizations, many of them associated with the FMLN, have been permitted to receive external funds through the PRN, under the auspices of Catholic Relief Services. The government and the FMLN also recently signed a joint memorandum of understanding with the International Fund for Agricultural Development (IFAD) where they agreed that government agencies and NGOs would work together in an IFAD project which is being negotiated for Chalatenango.

These are, however, still first steps. As long as important sections of the Salvadorean government distrust NGOs as leftist the change in NGO–state relations will not have been definitive.

Source: PREIS 1992

National Liberation Front (FMLN) and foreign donors, explicitly recognizes a major role for NGOs in the reconstruction process.

Another factor in improving relations between NGOs and the public sector has been the reduction of USAID support for NGO activities and the rise in the support they receive from social funds supported by the World Bank and the IDB (Reuben 1992). Whereas USAID encouraged NGOs to operate independently from the public sector, the multilateral banks are channelling their funds for NGOs through the public sector and are explicitly promoting better relations between the two groups.

A final element which is key to understanding the recent evolution of the NGOs and their relations with public agencies is the rising international concern with the environment (Box 8.2). International NGOs such as the International Union for Nature Conservancy and the Worldwide Fund for Nature were among the first to promote national systems of protected areas, and hence established early on the legitimacy of NGO participation in conservation efforts. Most public agencies involved with natural resources in Central America now have substantial experience in dealing with NGOs. Over time, the focus on natural resources has expanded from a narrow concern with protected areas to including broader issues related to agricultural sustainability – such as soil erosion, deforestation, pesticide abuse and loss of genetic diversity among plants used by farmers. This has led many environmental NGOs to work on agricultural issues for whom dialogue with the public sector is an established practice.

Box 8.2 Government–NGO collaboration in the International System of Protected Areas for Peace (Si-A-Paz)

During the late 1980s, the International Union for Nature Conservancy (IUCN) played a leading role in persuading the governments of Costa Rica and Nicaragua to develop a joint system to protect the tropical rainforests around the eastern portion of the San Juan river, which divides the two countries. The system includes a bi-national management structure responsible for managing the protected areas and contemplates activities related to forestry development, environmental education and basic studies. From the outset a number of national and international NGOs have been involved in this process. An action plan has been drafted which includes projects to be carried out by public national-resource institutions, universities, NGOs and municipal governments. The plan was recently presented jointly by these different institutions to a special meeting of international assistance agencies. It appears that all parties involved believe they have a better chance to obtain funds if they work together.

Source: IRENA/MIRENEM 1991.

THE NGO APPROACH TO AGRICULTURAL DEVELOPMENT

NGOs and agriculture

NGO interest in agricultural technology grew throughout the 1980s. Internal discussions and outside criticism led international NGOs that originally had a paternalistic approach, based on donations and subsidized credit, to shift towards projects designed to achieve sustainable improvements in rural incomes. NGOs set up to provide emergency support for refugees and displaced persons began to envision these problems as long-term and seek more permanent solutions. In the first part of the 1980s, donor enthusiasm led many NGOs to become involved with 'appropriate technology', although interest diminished in the latter part of the decade, due to the efforts' limited success. World Neighbors in Guatemala and Honduras generated interest in soil conservation among numerous NGOs. National NGOs which had their roots in the popular education tradition associated with Paolo Freire began to see agricultural technology as an important aspect to include in their adult education programmes. Large number of foreign agronomists came to the region, bringing with them methodologies and agricultural practices from throughout the world.

Given the great diversity among NGOs, it is difficult to talk about *an* NGO approach to agricultural development. Nevertheless, certain characteristics are common to most NGOs and there are some general trends in the sector as a whole.

Two common characteristics of NGOs with agricultural programmes are: a focus on the micro (community) level and a holistic view of human development, in which technological change is only one element. The emphasis on the micro level is partially a function of the NGOs' own limited geographical coverage. In addition, their relative isolation from national policy-making circles and the importance they attribute to grass–roots involvement has led them to concentrate on micro-level activities. Many national NGOs are aware of the limitations of a purely local approach, given that key variables such as prices, wage rates and access to land are largely determined at the national level. Nevertheless, even when they have been closely affiliated with national political parties or farmers' organizations, the NGOs involved in rural development tend to concentrate their day-to-day efforts on achieving marginal improvements in living standards in particular communities. There are some social-science research NGOs which concentrate on studying national policies, but few of them have agricultural programmes.

The NGOs' holistic approach leads them to include a wider variety of activities than their public-sector counterparts. For example, a majority of NGOs with agricultural programmes in El Salvador and Honduras provide social services such as health care and education, and are involved in community organizing (Kaimowitz *et al.* 1992; Rodriguez 1992). Many NGOs complement their promotion of new agricultural practices with credit, marketing assistance, and efforts to facilitate access to agricultural inputs. They also tend to incorporate women and children, rather than just male farmers.

Farmer participation

Almost all NGOs claim to have a participatory philosophy that actively incorporates beneficiaries' desires, knowledge and resources. Actual participation has generally been weaker than the rhetoric would imply. Although NGOs frequently hold meetings or assemblies to discuss plans for projects with their beneficiaries or interview them about specific topics, most decisions are still made by NGO managers and donor agencies, with little substantive input from their clientèle. Consultations take place after projects are under way, when key decisions about resource allocations have already been made.

One particular participatory practice which the NGOs use extensively is to employ farmers as paratechnical extension agents. More than half the NGO extension agents in El Salvador and Honduras are farmers (Rodriguez 1992; Kaimowitz *et al.* 1992). These paraprofessionals come from and live in the local communities, are able to communicate farmer sentiments to the NGOs, and know how to explain new methodologies and practices to farmers in terms they are familiar with (Box 8.3).

Box 8.3 NGO experiences with paratechnical extension agents

As with several other methodological innovations, the use of farmers as paratechnical extension agents in Central America was probably first started by World Neighbors in the 1970s. Since then the practice has spread widely and has even been adopted by public-sector institutions in several countries.

Typically, NGOs select farmers to be paratechnicians who have previously been participants in the organization's activities. Religious NGOs use people who have been leaders in the Catholic Base Communities or local Protestant churches.

Most paratechnicians continue to work in their own communities, although this is not always the case. Compensation varies from occasional support for food and lodging during courses to full salaries, normally at or below minimum wage. Some initial training is generally provided, as well as regular short sessions where new agricultural practices and logistical issues are discussed and programme progress is evaluated.

Besides being a potential vehicle of farmer participation, the use of paratechnical extension agents greatly reduces costs. Not only are paratechnicians' salaries much lower then traditional extension agents, they do not require transportation or daily allowances to work in the communities.

Despite their emphasis on farmer organization, the NGOs' relations with independent farmer organizations have often been problematic. Farmer organizations in Costa Rica have publicly complained that the NGOs' activities are largely self-servicing and carried out without consultation. Only a handful of NGOs in Honduras and Nicaragua have been able to work effectively with farmer organizations. Most of these were specifically established to provide technical assistance to farmers' groups affiliated to national peasant organizations which emerged from the agrarian reforms. NGOs have had better relations with farmers' organizations in El Salvador, largely due to the central role of the Christian Democratic Party and the FMLN in both types of organization, which has permitted greater coordination. Throughout the region, many farmers' groups themselves increasingly provide credit, technical assistance and other services to their members and see the NGOs as competing for donor resources.

The NGOs with the most beneficiary participation seem to be those directly associated with churches, political parties, and farmers' organizations. Even when these institutions are operated hierarchically, from the top–down, they tend to elicit a sense of belonging and loyalty from beneficiaries and a level of active involvement which is absent in most fully autonomous NGOs.

NGOs and agricultural technology

Several distinct approaches to agricultural technology can be identified among the NGOs. As in the case of the South American NGOs, these approaches are not mutually exclusive – rather it is that different NGOs tend to give differing emphasis to them in their work. To the extent that the following discussion presents stereotypical images, this is for ease of presentation. Debates continue within the NGOs.

The first, 'modernist' approach tends to view technology as synonymous with capital inputs (Box 8.4). Efforts centre on making seeds, plants, fertilizers, pesticides and grain storage containers available to farmers through credit programmes operated by the NGOs. Technical assistance consists largely of visits to individual farmers that typically lead to some recommendation regarding input use. This approach continues to predominate in many NGOs, but has declined in importance due to problems with loan repayment, concerns about creating dependency, and criticisms regarding excessive agrochemical use. NGOs that use this approach have tended to replace credit for individual farmers with group credit and to incorporate certain soil-conservation techniques in their programmes.

A second approach, often used to complement the first, concentrates on promoting new products, such as vegetables, fruits, medicinal plants, goats, pigs, milk cows and fish, to supplement farmers' income or for household consumption. Typically the NGO provides the initial seeds, plants or animals, as well as training and technical assistance. Most projects for women fall into this group. Common problems include limited markets for and experience with consuming the new products, dependence on continued NGO presence to obtain the necessary inputs, and lack of adequate household participation in selecting alternatives which respond to felt needs (Kaimowitz *et al.* 1992). The NGOs have played only a very marginal role in promoting small farmers' participation in non-traditional agricultural exports.

A third, 'agroecological' approach concentrates on substituting local resources for imported inputs and conserving natural resources. This approach, associated with terms such as sustainable agriculture, traditional agriculture, and organic farming, is rapidly gaining adherents among NGOs. The recent rise in real agricultural-input prices has made NGOs particularly conscious of the need to find cheaper alternatives to purchased inputs. To date, most efforts have gone into organic fertilizers and green manures, mechanical soil conservation structures and the use of natural repellents for pest control (IDESAC *et al.* 1989). Interest in agroforestry and more sophisticated pest-management techniques is also growing. Groups that work with these practices tend to emphasize farmer motivation and training rather than agricultural credit or individual visits.

Perhaps the main limitation with this approach as applied by NGOs has been a tendency to promote the same few simple practices in a wide variety of contexts and with farmers who have different needs, access to resources and agroecological conditions. Often these practices are presented as if they were universal panaceas, which they are not. This same anti-scientific approach has led some NGOs to be overly dogmatic in their rejection of purchased inputs, even when these may be justified and farmers are interested in using them.

Box 8.4 NGOs with a modernist approach to agricultural technology

The NGOs in this category can largely be divided into two groups: large international NGOs which developed to provide charity, and smaller national NGOs which have focused their technological efforts on providing agricultural credit. The first group includes Catholic Relief Services, Save the Children, Plan International, World Relief and World Vision. They are major international institutions, with large numbers of employees in each country. Many are beginning to abandon a pure modernist approach, and incorporate more ecological and participatory elements. The second group includes the majority of NGOs in the region, which have formed small agricultural components by hiring one to five agronomists and providing them with operating funds. In the latter case, the agricultural efforts tend to lack a clear methodology. The agronomists find it easier to work with farmers if they can offer credit or small donations of seed and inputs.

A fourth, 'alternative development' approach focuses more on process than on specific agricultural practices (Box 8.5). These processes include popular education, farmer organization, farmer experimentation and participatory diagnosis. The fundamental premise is that rather than providing farmers with new technologies, the primary role of the NGOs should be to improve farmers' capacity to analyse their context, generate and select technologies, work collectively, manage their resources and present their needs to external agencies.

A strong emphasis on process is absolutely essential to achieve sustained improvements in local conditions. Care must be taken, however, not to let process become an end in itself rather than a means. If, after a time, these methodologies do not lead to concrete improvements, farmers can become tired of participating and begin to withdraw. Much of the early efforts of the NGOs concerned with process centred around persuading farmers to produce collectively, with limited success. Few professionals in Central America still actively promote fully collective production.

More recently, greater attention has been given to farmer experimentation. NGOs such as the Centre for Research, Promotion and Rural and Social Development (CIPRES) in Nicaragua and the Western Regional Rural Development Association (ADRO) in Honduras have urged farmer collaborators to design their own informal experiments with ideas or materials provided by the NGO. The farmers comment on what they observed during the trials and share the results with their neighbours.

The sources used to learn about proposed agricultural practices have varied depending on the type of technology involved. Most information regarding agricultural-input use comes from the NGO agronomists' initial training, their prior experience working in the public sector or public-sector publications. In Honduras two public-sector projects financed by the Swiss, POSTCOSECHA and PROMECH, have been key in promoting grain storage techniques and modified agricultural implements (Kaimowitz *et al.* 1992). Information regarding new production alternatives comes from both public and private sources. In several countries the Heifer project has been an important source of information about raising small animals and cows.

Box 8.5 NGOs with an alternative development approach

NGOs with an alternative development approach tend to have roots in the cooperative movements or in popular education. Typical examples of the first group include the Honduran Institute for Rural Development (IHDER), the Research and Training Institute for Cooperatives (IFC) and Canadian University Services Overseas (CUSO) in Honduras. The second group includes centres such as ALFORJA in Costa Rica and Cantera, CIPRES, Nitlapan, Xezontle and the Centre for Education and Agricultural Promotion (CEPA) in Nicaragua. For many of these NGOs agricultural technology is a means towards consciousness-raising or political mobilization rather than a goal in itself.

Almost all the information about low-input and resource-conservation technologies has come from the NGOs themselves. As mentioned previously, World Neighbors has had a major influence in promoting soil-conservation techniques. International NGOs which operate in numerous countries such as CARE, World Vision, Save the Children, Plan International, Mennonite Central Committee and Catholic Relief Services actively exchange information between their national offices. Occasionally this information is made available to national NGOs operating in the same countries. Despite much rhetoric to the contrary, practices promoted by NGOs rarely come from farmers.

In recent years, efforts have increased to improve information exchange about agricultural technology between NGOs. Bread for the World has sponsored three regional meetings of NGOs from Central America and Mexico involved with alternative technologies and has financed the preparation of a series of national directories of institutions working with alternative technologies. A Swiss Development Cooperation (COSUDE) project which works with NGOs involved in agriculture in Honduras has expanded its activities to include all of Central America and is promoting exchanges between NGOs in the different countries. In Costa Rica, a group which coordinates NGOs involved in alternative development projects (COPROALDE) has held two regional conferences on traditional agricultural practices.

Almost all the NGOs restrict their participation in the Agricultural Technology System (ATS) to the transfer sub-system. Few are involved in research. The limited research which they do is almost entirely adaptive research. NGOs with research activities include: CEMAT in Guatemala, the Rural Development Centre (CEDRO) in El Salvador, the Agricultural Centre for Research, Development and Training (PRODESSA) in Nicaragua, the International Centre for Cover Crop Documentation (CIDICCO) in Honduras, the Educational Corporation for Development in Costa Rica (CEDECO), and several others. Perhaps twenty or thirty other NGOs have informal trials to test different agricultural practices. The remaining NGOs must rely on outside sources of technology for the practices they promoted, and these sources frequently have few relevant technologies to offer.

Another source of technical weakness in the NGOs is the composition of their personnel. Rarely do NGOs have staff with postgraduate degrees. In fact more than two thirds of those involved in agricultural programmes in El Salvador, Guatemala and Honduras have no university degree (Rodriguez 1992; Matheu 1992, Kaimowitz *et al.* 1992). The remainder have only secondary-level technical training or are farmer paraprofessionals. This staff composition has significant advantages in terms of costs and the staff's ability to work closely with farmers, but it makes it difficult for NGOs to search for and assess new technical options, to understand fully their potential and limitations, and to work effectively with more complex technologies, such as certain practices involving agroforestry, integrated pest control and irrigation techniques. In comparison with the public-sector institutions, NGOs have significantly more expatriate professionals, social scientists and paratechnical staff, but fewer national agronomists with degree level or postgraduate training.

LINKS BETWEEN NGOs AND THE PUBLIC SECTOR

Factors favouring links

The current need and potential for NGO–public-sector collaboration to develop and promote relevant agricultural technologies is greater than ever. The decline in public-sector resources and credibility makes it very difficult for public institutions to achieve major technological changes in small-farm agriculture without the support of NGOs and other private institutions. NGOs need support from public-sector professionals and international research centres to improve their technical capacity. The reduction of ideological polarization has reduced the political barriers which separated NGOs and governments and raised the possibility of incorporating NGOs into broader policy debates.

One major factor promoting NGO–public-sector links has been donor agency support for joint activities. As noted previously, the World Bank has actively sought NGO participation in the social compensation programmes it finances which are managed by national governments. Thus, for example, through the Honduran Fund for Social Investment (FHIS), financed by the World Bank, the Honduran public sector has contracted the services of NGOs to establish household gardens or small animal projects (Kaimowitz *et al.* 1992). Through its Central American Basic Grain Research Project, the European Economic Community has also promoted public-sector–NGO coordination in agricultural research and technology transfer. As noted previously, in El Salvador various donor agencies plan to finance joint NGO–public-sector projects as part of the National Reconstruction Plan which emerged from the peace accords (PREIS 1992).

Donor interest has forced Central American governments to take the NGOs more seriously. As a result, most governments have created special offices for government–NGO relations, often directly under the Presidents. They have also collaborated with NGOs to seek donor support for specific projects. Faced with difficulties in obtaining donor support for public-sector initiatives, the Honduran government is now attempting to channel donor resources towards NGOs supported by the ruling National Party.

Political factors have been crucial to improved relations in El Salvador and Nicaragua. For example, during a recent joint seminar between the public research and extension agency (CENTA) and NGOs in El Salvador it was evident that both groups made a special effort to find points of agreement, at least in part as an expression of goodwill following the national peace accords. Efforts by Nicaragua's National Centre for Basic Grain Research (CNIGB) to work with several NGOs have also been influenced by the Chamorro government's conciliatory posture towards Sandinista sympathizers.

A third factor favouring public-sector–NGO relations is the close personal ties between many of the people working in the two groups. A majority of NGO staff in Nicaragua, and a significant minority in the other countries, formerly worked in government agencies. These individuals are familiar with public-sector procedures and maintain friendships with public-sector agronomists. It has even become common for government employees to supplement their incomes through part-time participation in NGO activities. On the local level, NGO and public-sector personnel frequently come into contact and seek to find ways to assist each other.

In El Salvador and Honduras, government officials increasingly see collaboration with NGOs as a way to reduce expenditure on public extension services, and to support institutions which are more efficient at providing services. This fits in well with the governments' overall interest in privatization. The national legislature of Honduras has even gone so far as to pass recently a law requiring the government to elaborate a plan for privatizing agricultural research and technology-transfer activities, which takes into account NGOs as well as private firms and farmer organizations. For the most part, however, this factor has played only a secondary role in the actual links which have been created. The short-term political, personal or economic interests of the institutions involved have been more important. Few people in either the public sector or the NGOs have really begun to think about what the most appropriate long-term relation between the two groups might be.

Obstacles to links

Even with these favourable tendencies, strong obstacles remain to NGO–public sector coordination. These include the institutional weakness of the public sector, differences in clientèle and approaches, the dispersed nature of the NGOs, and continuing distrust and competition.

Many public institutions are currently so unstable that it is practically impossible for them to guarantee that they will fulfil commitments made to NGOs. Good working relations are often achieved at the local or regional levels, but come to an end when the individuals involved resign or are transferred. Public research and extension agencies cannot commit themselves to joint visits or trials when funding for these activities is uncertain. The decline in public-sector research and the desertion of many of the governments' best professionals severely limits the possibility of providing NGOs with relevant recommendations or technical support.

Even though most public-sector agencies claim to focus on small and medium farmers, in fact they work mostly with the more commercial and better-endowed farmers in these groups. NGOs work more with farmers in less favourable agroecological environments and with marginal groups such as women and indigenous farmers. These differences affect the type of technologies emphasized by each group and limit the opportunities for productive interaction.

Most public-sector agronomists continue to promote chemical fertilizers and pesticides, which are increasingly rejected by the NGOs. Public-sector agronomists tend to feel that the NGOs' approach to technological problems is overly romantic and has little scientific basis. New varieties, which are at the centre of most public-sector research, are of only marginal interest to most NGOs. It has generally been easier for the public sector to work with NGOs which promote the use of agrochemicals than with those sympathetic to organic farming.

Whereas public agencies tend to operate at the national level, the NGOs are found mostly at the local level. This makes it hard for national research and extension directors to identify a few major NGO representatives with whom to interact. It can also lead national directors to underestimate the true magnitude of NGO efforts. Because decision-making in public institutions tends to be rather centralized, regional directors and local researchers and extension workers rarely have the authority to commit resources to working with NGOs. Moreover, if anything, the trend in most Central American public research and extension institutions is towards greater, not less, centralization.

In Costa Rica and Guatemala, coordination between the major public institutions and the NGOs has been practically impossible. In Costa Rica, the Ministry of Agriculture and Livestock (MAG), which is responsible for research and extension, does not feel the need to coordinate with NGOs, because it perceives them (somewhat mistakenly) not to be important. In Guatemala, political distrust between the NGOs and the government, and the general climate of violence in the country, continue to limit cooperation. There also continues to be strong, ideologically motivated opposition to links with NGOs from certain sectors within the governments of El Salvador, Honduras and Nicaragua.

Linkage mechanisms

Despite the recent and limited nature of NGO–public-sector coordination, a number of different structural and operational linkage mechanisms have been tried. One structural mechanism is government funding of NGO activities and vice versa. Government funds for NGOs have largely been restricted to the social-compensation programmes mentioned previously, the agricultural components of which it is still too early to evaluate. The most notable case of NGO funding for government activities is in Nicaragua, where CARE has contracted the National Centre for Basic Grains Research (CNIGB) to conduct agronomic trials and provide seeds. This has permitted an effective complementarity, since CARE lacks research and seed production capacity and CNIGB needs operating funds.

In other cases, donor projects have simultaneously funded public-sector and NGO activities. USAID, for example, supported both public-sector and NGOs' extension efforts in its Rural Technologies and Natural Resource Management Projects in Honduras. The EEC Basic Grains Project mentioned several times above also funds both NGOs and public institutions. These situations have stimulated communication between the two groups, even when the activities are largely planned and executed independently.

As noted above, progress has just begun towards developing more long-term structural relations between the public institutions and the NGOs. The Salvadorean government has begun to discuss the possibility of dividing the country into areas where NGOs would be responsible for agricultural extension and others where the public sector would operate (IICA *et al.* 1992). A handful of NGOs participate in the national councils established by the Central American Basic Grain Research Programme to approve project proposals for funding.

Operational linkage mechanisms which have been tried or proposed include staff secondment, joint planning exercises, joint activities and the exchange of materials and information. The possibility of seconding government agronomists to work with NGOs has been proposed in Honduras as a means to reduce public employment, but has still not been implemented. There have already been instances where the government has seconded extensionists to work for farmer organizations. During the first few years the government continues to pay the agronomists. After that the farmer organizations or NGOs would have to assume the costs.

Relatively few planning exercises have involved both public agencies and NGOs. Undoubtedly the most important has been the negotiations surrounding the National Reconstruction Plan in El Salvador. NGOs and public natural research institutions have worked together to formulate proposals for protected areas, including the agricultural activities planned for the surrounding buffer zones, in several countries.

Joint activities between NGOs and public agencies are increasingly common. The POSTCOSECHA and PROMECH projects mentioned previously operate out of the Honduran Ministry of Agriculture, but provide training on grain storage and agricultural implements to a wide range of public and private agencies (Kaimowitz *et al.* 1992). They also assist in making materials available to produce grain silos. In the POSTCOSECHA project, each NGO is requested to designate a liaison person responsible for contact with the project and there is an annual meeting of all the institutions involved to evaluate the project's progress and discuss possible improvements. Honduras Professional Training Institute (INFOP) has also offered a number of courses at the request of NGOs. In Nicaragua, the National Basic Grains Research Centre has organized joint events with several NGOs related to farmer experimentation. Join activities such as these, where the content and institutional responsibilities are clearly defined and focus on a couple of specific topics of mutual interest, allow all parties to benefit, without limiting institutional autonomy. Seminars have been held in all five Central American countries at which both public-sector and NGO representatives have presented their experiences. As yet, none has led to the formulation of joint activities, but they have been important fora for exchanging information and improving social relations between the two groups.

More limited, but important, forms of collaboration include NGO usage of public-sector publications, purchase of seeds, plants and animals from public agencies, and informal coordination at the local level. Unfortunately, these forms of collaboration have tended to diminish with the decline of public-sector presence.

Finally, it is worth noting that NGO collaboration with international and regional research institutions such as the Research and Training Centre for Tropical Agronomy (CATIE), International Centre for Tropical Agriculture (CIAT), Organization of Tropical Studies (OTS), International Centre for Maize and Wheat Improvement (CIMMYT), and the Panamerican Agricultural School (Zamorano) has also increased substantially. The principal motivation in these relations has been the NGOs' desire to obtain training for their staff and access to funds. For several years CATIE and OTS have been training NGO personnel in forestry, agroforestry and agroecology. NGOs have also been the principal counterparts of CATIE's 'Conservation for Sustainable Development Project in Central America' (Krantz *et al.* 1992). Zamorano recently began a major project to train NGO paraprofessionals in Honduras in integrated pest management. As part of the project, the paraprofessionals are exposed to basic concepts of insect biology and actively encouraged to use those concepts to experiment and develop new pest management techniques. Substantial funds are being allocated to monitor the experience of these paraprofessionals after they return to their communities. The other international centres have just begun to work with NGOs, but have already taken a number of important steps in that direction (Box 8.6).

Box 8.6 Major types of linkage mechanisms between public institutions and NGOs in Central America

Structural:

- Public-sector contracts NGO services or vice versa
- Common funding source supervises both groups
- Agreed division of labour
- Joint committees to allocate resources

Operational:

- Staff secondment
- Joint planning exercises
- Joint activities
- Interchange of materials
- Communications (seminars, publications, etc.)

Assessment of the linkages

In principal, public-sector–NGO linkages should help the Agricultural Technology System to provide more efficient, client oriented services. Most of the applied and adaptive research sub-system is still largely concentrated in the public sector, while the NGOs have proven that at least in certain instances they can be relatively efficient and more responsive vehicles for technology transfer to small farmers. The majority of NGOs are basically implementers. They lack capacity to make institutional innovations, do research or provide adequate feedback to the national agricultural research system. There is, however, a small but significant group of NGOs with capacity in these areas, which could make an important contribution to the overall functioning of the Agricultural Technology System.

For the most part, to date, few people in either the public sector or the NGOs have given much serious thought to the division of labour and links that would be required to achieve a more effective, efficient and responsive technology system. The links which have occurred respond more to short-term institutional and political considerations.

THE PROSPECTS FOR THE FUTURE

Recent progress in resolving the military conflicts in Central America has opened the way for greater collaboration between NGOs and the public sector, but it also poses major dangers. The most important is that foreign donors will lose interest in the region and reduce their support for both the NGOs and public institutions. Already there are signs this is occurring. If it does, neither the public sector nor the NGOs will have much to offer the other, and little progress is likely in improving small-farm technology. It is in both groups' interest to have strong counterparts.

The specific situation differs in each country. The most promising outlook, at least in the short run, is in El Salvador. As a result of the peace accords, substantial new resources have become available both to the public sector and the NGOs. A national consensus has emerged regarding the need to reduce rural poverty and there is a new climate of goodwill, which is propitious for collaboration.

In Nicaragua and Honduras there are encouraging signs of greater inter-institutional coordination. However, foreign assistance is declining, and public agricultural agencies are on the verge of collapse. War and violence continues to plague Guatemala. Costa Rica's public-sector agencies and NGOs have yet to acknowledge each other's existence.

Natural-resource conservation and social compensation are two areas where donor support will continue to be strong. Both are good candidates for greater NGO–public-sector interaction. Under pressure from foreign donors and national ecology movements, public research and extension institutions may move closer to the NGOs' pro-environmental stance, and become more involved with resource-conserving, low-input technologies. The difficulties in generating massive environmentally friendly technological changes should not be underestimated, however, and this may eventually lead to frustration. Many factors discourage the use of these technologies and much remains unknown about their implications.

The social compensation programmes financed by the multilateral banks are likely to grow and may convert the NGOs into being solely service providers. This could undermine the NGOs' holistic character and concern for process, and lead them to lose their creativity and dynamism and become merely low-quality subcontractors for the state.

The relations between NGOs and international research centres are particularly important because with the decline of national public-sector institutions, the international centres are now among the few institutions with sufficient technical expertise and resources to support NGO activities effectively. The international centres can also play a major role in promoting greater ties between NGOs and national public institutions.

The new relations between international centres and NGOs pose difficult questions about how much international centres should become directly involved in field-level development activities. It has also already begun to promote certain institutional jealousies on the part of the public-sector institutions, which are the centres' traditional counterparts, and fear competition from the NGOs.

The challenge faced by NGOs, public-sector institutions, and international research centres in Central American small-farm agriculture is greater than ever. The restoration of peace in the region could lead to higher and more equitable agricultural growth, but there is certainly no guarantee that it will. The agricultural frontier is reaching its limits and destroying much of the region's remaining primary tropical forests in the process. If the region is to avoid increasing dependence on imported foodstuffs, it must raise yields, and do so in a context of increasing input prices and a deteriorating natural resource base. Most small farmers still have no viable alternatives outside of agriculture. The Central American peace process has opened the doors for collaborative efforts to solve these problems. A failure to take advantage of that opportunity can only lead to a return to the violence and social unrest from which the region is just beginning to emerge.

NOTE

1 Most of the chapter's material comes from national studies of NGOs with agricultural programmes in El Salvador, Guatemala, Honduras and Nicaragua carried out in 1991 and 1992 by Roberto Rodriguez in El Salvador, Roberto Matheu in Guatemala, Byron Miranda and Pedro Ugarte in Nicaragua, and David Kaimowitz, David Erazo, Aminta Navarro and Moises Mejía in Honduras. These studies were financed by the Central American Basic Grains Research Programme (PRIAG), the United States Agency for International Development (USAID) and the Interamerican Institute for Cooperation in Agriculture (IICA). Anthony Bebbington and Manuel Chiriboga made useful suggestions, which have been incorporated into the chapter.

9

CONCLUSIONS: RETHINKING ROLES FOR NGOs AND THE STATE

Challenges of a new agenda for agricultural development

We began this book with a question. Can rural development strategies in Latin America be reoriented so that they generate agricultural technologies that are more appropriate for small farmers, enhance the capacity of the rural poor to make use of those technologies and increase their influence over government agricultural development institutions? We suggested that non-membership NGOs can make important contributions to this reorientation.

The ground we have covered since posing this question suggests that agricultural development strategies are currently undergoing quite fundamental changes that might favour such a reorientation. From a past in which strategies focused on the transfer of Green Revolution technology to the rural poor has emerged a present in which even orthodox agencies question how appropriate such strategies are for the rural poor. From a past in which roles were relatively well defined for both NARS and NGOs has emerged a present in which they are in question and are being restructured.

These changes increase the opportunity for NGOs to influence programmes of ATD so that they are more participatory and more appropriate for the rural poor. NGOs will do so in different ways: by working with the state in the design and implementation of agricultural development programmes to enhance their effectiveness; by interacting with the state to challenge the thinking behind those programmes; and by building capacity at the grassroots so that the rural poor themselves exert pressure over the state. For many NGOs, these interactions with the state imply a change from strategies they have pursued in the past. This change will exert tensions and strains in the organization. However, we have suggested that the traditional tendency of NGOs to work in isolation and avoid contact with the state is increasingly untenable.

To conclude the argument we have developed in the book, in this chapter we review the recent fundamental rethinking of the technological and institutional basis of agricultural development that is forcing the issue of NGO–state linkages on to the agenda. We then consider how far NGOs will be able, and willing, to respond to the challenges presented to them by this restructuring of agricultural development. We then synthesize our conclusions on how more-intensified forms of interaction might allow NGOs to improve the quality of their work and widen their impact by engaging in policy debates with the state. We close with reflections and recommendations for the main actors in these changes: donors, public-sector officials and NGO managers.

RETHINKING AGRICULTURAL DEVELOPMENT IN LATIN AMERICA

Changing institutional roles in Latin American development

Until the 1980s the roles of the state and NGOs in Latin American development were more clearly defined. In the dominant development strategy of growth grounded in modernization, the state's role was to intervene, assist and often direct the form this development took. In the agricultural sector this meant land reforms, rural development programmes and the creation of an institutional structure, the NARS, whose role was to be the central actor in generating and disseminating modern technologies. The NGOs, to those who knew them, had an equally well-defined set of roles: to be allies of the poor, to increase their access to some of the fruits of growth, to contest much of the project of modernization, to criticize the forms in which the state intervened and to be voices for a more democratic form of development. These were the roles of, and relationships between, NGOs and the state that we sketched in Chapter 3. It is recognition of these roles that has inspired one dimension of the current interest in NGOs: that which sees them as the vanguard of a more democratized, participatory and equitable rural development.

A series of sociopolitical and economic changes has, however, called these prior roles into question. Neo-liberalism and the fiscal crisis of the state have left the contemporary public sector smaller, weaker and unable to intervene as it has done in the past. Furthermore, there is a policy commitment among donors and governments that it ought not do so. The state is receding and is looking for new partners to help implement and fund social and development programmes. The social funds have been the clearest example of such government delegation and subcontracting of social and development work to the private and non-governmental sectors.

These changes have influenced thinking among those activists who seek an alternative development and who have spent much effort calling on the state to intervene more on behalf of the poor. Many are slowly coming to accept elements of the neo-liberal argument that the public and centralized institutions have inherent inefficiencies. More significantly, some seem to acknowledge that some of the perspectives of neo-liberal resonate with much of what they have long been saying about grassroots, bottom-up development:

that popular sectors should be given greater control over their livelihoods and over local development initiatives. Of course, neo-liberalism is not the same as grassroots development – the latter would still expect that there should be more protective policy environments for, and redistribution of surplus to, these popular sectors as they develop their own local initiatives. None the less, there is a shared ground: that the state should be less involved in intervention, and civil society (including the NGOs) more so.

Other significant changes for the relationship between NGOs and the state stem from the fact that Latin American politics has moved on from the years of repressive dictatorships. However uneven and faltering the process of democratization may have been, it has led to changes that give the state less of a free hand to repress and harass NGOs (see Ch. 3). Furthermore, the democratization process has been linked to processes of decentralization, and to donor pressures to enhance the transparency and flexibility of public administration. These changes present NGOs with a new challenge: now they, as unelected institutions, are faced with governments that have been elected. But they also present to NGOs the possibility of participating in policy discussions in an unprecedented way.

The politics of civil society has also moved on. NGOs are now far more heterogeneous than they ever were before: the radicals have been joined by the reformists, and the ideologues by the opportunists. Many NGOs are driven today less by ideals than by the income-enhancing concerns of a professional middle-class feeling the pinch of structural adjustment. These sorts of NGOs would be more than willing to fill gaps left by a receding state – as long as they are paid to do so. Even some of the more 'politically correct' of NGOs no longer reject these proposals purely on the basis of ideology. If funds are available from public-sector and donor budgets, an increasing number of NGOs find themselves able to accept this finance.

Whilst NGOs and the state still have much to iron out and negotiate in this new context, what is clear is that their former roles and dominantly conflictive relationships are no longer wholly relevant. In short, a series of sociopolitical and economic changes have led to a situation in which there is a far greater possibility that NGOs and the state will enter into more direct and often collaborative relationships in development activities – a possibility that would in most cases have been unthinkable in the past.

A new technological agenda for agricultural development

The basis to traditional strategies of ATD in which NARS were a central actor was to increase the use of high-yielding varieties, intensify agrochemical-input use and introduce mechanization into an increasing array of farm operations. This was done through centralized research on experiment stations and the transfer of technological packets through extension services.

This strategy is now being questioned. Is it ecologically sustainable? Are we seeing declining returns to the Green Revolution approach to technology generation? Is it equitable? Is it a form of cultural domination? Can modern technology address problems of rural poverty, particularly when technical assistance is not accompanied by other forms of rural support? Many institutions are asking these questions, although their motivations are not all the same. There are, for instance, quite different politics behind an International Agricultural Research Centre 'going green' and an indigenous farmer organization looking for a development grounded in traditional practices. However, they stem from the same uncertainty about the rural development model that has held sway for decades.

This questioning has given rise to outlines of new proposals for agricultural development for the rural poor; we discussed some of these in Chapter 4. These proposals have both technological and institutional implications. Some of them try to rework elements of the traditional Green Revolution strategies, seeking ways to close technology gaps, to increase the access of small farmers to new technology and to render that technology more adapted to their circumstances. Others, far more critical of modernization, propose new approaches grounded in agroecology, indigenous knowledge and locationally-specific technology generation.

Despite their differing emphases, there is a clear middle ground to these proposals. This middle ground for a new technological agenda is characterized by a set of concerns for lower-input technologies, decentralized institutional configurations to allow the adaptation of technologies to local conditions, and the provision of training and material support to help farmers make more effective use of new technology. Similarly, there is a shared concern to find mechanisms to increase rural incomes through interventions in the food system beyond the farm gate, particularly in marketing and processing spheres. Overall, in our adopted idiom of the ATS, there is growing concern to strengthen actions in the sub-systems corresponding to adaptive research and the transfer and use of technology, and to identify appropriate forms of institutional interventions in these sub-systems, above all at a local level.

This emerging technological agenda has generated interest in NGOs for several reasons. Many suppose that NGOs are more inclined to work with low-input technologies than the NARS. Similarly their grassroots orientation inspires claims that NGOs would be better able to adapt technologies to local conditions, create a bridge between local campesino knowledge and external expertise, and provide the additional forms of support that campesinos need if they are to benefit from introduced technology. We considered these claims in Chapter 5. That review of NGO experiences suggested that they do indeed operate with a more systemic perspective on farmer resource use, a more participatory approach to ATD, and have a more flexible and locally responsive institutional structure. They also have a greater proclivity to meld agroecological and modernizing perspectives than does the public sector. While there are extreme 'green' NGOs, many are located in the middle ground of the proposals emerging for the next steps beyond the paradigm of agricultural modernization.

At the same time it is widely recognized that government research and extension services simply lack the resources to meet this challenge and provide the intensity of coverage required. From this realization among donors and governments that gaps must be filled, and this gap-filling coordinated, has come the turn to NGOs. They have been invited to participate more actively in government coordinated and monitored programmes. If the state cannot respond to the challenge of a more agroecological, participatory and appropriate process of technology development, and if it is unable to give farmers the support they require at a local level – the argument goes – then the NGOs can replace it in this role, especially given that this sort of development is their forte. This instrumentalist attitude to NGOs is quite different from the interest in their role as agents of democratization.

Here the thinking on a new technological agenda converges with the other political and economic trends in Latin America that cast a reduced role for the state in implementing and bearing responsibility for development programmes, and an expanded role for 'the institutions of civil society', among them the NGOs. As we saw in Chapter 6, proposals for dividing up tasks in agricultural technology development between NGOs and NARS are now quite common, and the recent flurry of institutional changes in the public sector to operationalize such proposals has been remarkable.

RESPONDING TO THE CHALLENGE: NGOs AND THE NEW AGENDAS FOR AGRICULTURAL DEVELOPMENT

In several parts of the book we have commented that most government and multilateral donors' proposals for NGO involvement in ATD have approached NGOs with a highly instrumentalist attitude. They have expected NGOs to fill gaps left by the withdrawal of the state. Will NGOs accept such a role? Indeed, how able are they to do the magic that many donors and governments seem to expect from them?

The new breed of opportunistic NGOs might happily accept such a role. They are the NGOs we can expect to receive the state and donor funds channelled through NGOs as part of programmes of public-sector reform (i.e. cutback). To the extent that such NGOs are more efficient and flexible in their operations than state agencies, this in itself may make these programmes more effective. However, the more such funds are channelled through these sorts of NGOs, the more the label of NGO is devalued, as society itself recognizes that creating non-governmental organizations is becoming an income generating strategy for the middle class.

As for the more progressive NGOs, there are still reasons why many of them would reject such an explicit, and extensive, 'gap-filling' role, despite the fact that NGO–government linkages now seem more feasible than they were in the past. These reasons have much to do with NGOs' histories. Such 'gap-filling' would bring these NGOs into contact with a state that in earlier periods has proven itself to be unreliable at best, and dangerous at worst, and some of those NGOs that experienced the consequences of such repression still have serious reservations about getting too close to the state. This may be particularly so in Central America.

A more significant reason for these older NGOs to question such instrumentalist proposals lies in their own long-held concept of the state. In their view, the state should acknowledge the existence of a social contract in which society has a debt to its poorer strata in repayment for the burden of recessionary policies that they have carried, and for the surplus extracted from them in contemporary and historic periods.[1] For such NGOs, it is difficult to accept a role in which they are ostensibly replacing the state in fulfilling this function, and so contribute to the reversal of a concept of societal responsibility that has previously been at the core both of rural development policy and of NGO activities. Even if they are in the process of rethinking their conception of state–society relations, they are not willing to throw away the idea that society and the state owe a debt to the poor.

These seem to us very valid concerns – but they should not lead us to conclude that NGOs should reject contact with the state. Nor should they lead us to suppose that NGOs themselves will reject such relationships with the state. Our evidence suggests good reasons for believing that, while acknowledging the risks inherent in doing so, NGOs will take the challenge of linkages seriously. These reasons stem from evidence of:

- the grave limitations on the development work of NGOs when they act alone;
- the political weakness of NGOs when they isolate themselves from policy debates and decision-making within the state;
- NGOs' past practice.

The limitations of the NGO model

In Chapter 5 we concluded that on balance, the evidence is that NGOs of the type on which we have focused our attention have made real contributions to more participatory and appropriate forms of grassroots agricultural development. Yet we also suggested that this does not add up to a sufficient recipe for an alternative, appropriate and democratic development. There are still weaknesses in the NGO mode of operating.

First we suggested that there are constraints on the capacity of individual NGOs to operationalize their conceptions of participatory, systems-based ATD and to sustain their development initiatives. Their small size means they lack research facilities and professional/technical expertise. Though participatory, they too admit to failing on many occasions because of a poor understanding of local production systems and livelihood concerns.[2] The pressure to act quickly, often because of donor timetables, can and has lead to misspecified interventions.

Second, we argued that there are limitations on NGO capacity that stem from the structure of the NGO sector. (1) NGO presence is patchy – they do not cover the whole country and so *per se* cannot possibly be an alternative institutional vehicle for implementing a national programme of grassroots agricultural development (assuming there was one). (2) The lack of coordination among NGOs leads to inefficiencies poor communications and competition, which is itself aggravated by fights over funding. (3) As a consequence of their social origins, NGOs are still in many respects more akin to the state than to the grassroots. They may be closer to the rural poor than are public agencies, but their claims to be allies of the grassroots are being questioned by those they allegedly represent: an indication that their own actions would need to be democratized in order to enhance their contributions to the alternative development based on empowerment, or the democratic development that people like John Friedmann (1992) and David Lehmann (1990) expect them to carry forward.

The political costs of staying outside public-policy processes

NGOs are at present still too often outside policy debates for their visions to influence national programmes. Ultimately NGOs have to recognize that they are merely vehicles of development intervention at local level, whose outcome depends far more on a wider policy environment than the participatory methodologies they may have used in their projects.

This is part of a more general point: sustaining a grassroots agricultural development that alleviates poverty and protects the environment will depend on appropriate policy commitments. If NGOs are to play a role in identifying, implementing and enforcing these commitments, then some form of engagement with the state will be necessary. NGOs might be able to contribute to that through their perspectives and experiences, but their participation in programmes will still be dependent on those policies.

The reality of NGO past practice

There is a certain irony when NGOs' reject out-of-hand those proposals that expect them to fill the gaps left by the state – because in effect this is what NGOs have long since done. They have worked in places where the state was absent. They have worked with groups whose needs the state had left unattended. They have worked with methods and technologies ignored by the state. They have worked in sub-systems of the ATS in which the state has been weak or absent.

To date, then, NGOs may well have done more to deflect pressure from agricultural development programmes than they have done to democratize them. Of course, while they were filling gaps NGOs might have been at the same time empowering and organizing the rural poor in the process. But the bottom line is that they have been subsidizing public sector programmes for a long time. It was, for instance, the NGOs who responded to the social effects of Pinochet's adjustment policies, contributing to the diffusion of social tensions and so probably contributing to the survival of these programmes (Lehmann 1990).

In many respects, the recent proposals for public-sector retrenchment in agricultural development have therefore formalized a situation which already existed. Rather than complain about this, NGOs would be more consistent to the logic of their own approaches to development if they grasped this opportunity to try and change the situation by involving themselves more directly in arguments with the state and its donors about policy.

NGO past practice also shows us that in fact they have already engaged in linkages with the state which, while not always successful or easy, have not been as disastrous as some of the more extreme fears about the impacts of such linkages might lead us to have expected. These experiences suggested that the technical quality of NGOs' work has been enhanced by collaborative interactions with the state. The challenge is to consider how they might gain more. These experiences were discussed in Chapter 7 and are the basis of the discussion in the following section.

NGO-GOVERNMENT INTERACTION: ENHANCING NGO EFFECTIVENESS AND EXPLOITING ROOM FOR MANOEUVRE IN POLICY DEBATES

For the reasons outlined in the preceding section, we therefore suggest that the traditional roles and perspectives of NGOs will no longer do: they are often grounded more in past than in present realities; and they are themselves circumscribed (some would say greatly so) in the impact they can have on rural poverty. Much of their work has involved them in isolated gap-filling activities which have had no mechanisms to widen their impact beyond the locality, nor to draw down resources of other institutions to strengthen this local work. The NGO model is a very imperfect one.

We have therefore argued in this book that the evidence suggests that more intense *interaction* between NGOs and government institutions is desirable. Such interaction should not always be collaborative: NGOs have much criticizing left to do. None the less, successful critique requires contact and communication. So the second part of our conclusions deals with what form this interaction might take to best effect and what factors might facilitate it.

This interaction must be managed and structured in order to be effective in enhancing the quality of work that NGOs do as well as to increase their room for manoeuvring in making constructive criticisms of public programmes that fall on the ears of those who need to hear. In this section we summarize the evidence on the types of linkage we have identified in the case material, on their effectiveness and on ways in which this might be enhanced.

Interaction for complementarity: making linkages work

We used the terminology of sub-systems of the ATS, technology paths and linkage mechanisms in order to discuss interactions between NGOs and NARS aimed at exploiting and enhancing the complementarity of their work (Table 9.1). This terminology makes clear the different functions that NGOs and NARS play in sub-systems of the ATS, and the fact that linkage mechanisms can facilitate relations between institutions operating in different sub-systems or the same sub-system. Consequently, they can enhance the functioning of both their own work and of the ATS as a whole.

From this discussion, several patterns emerged:

- The work of stronger NGOs bridges the adaptive research and transfer sub-systems.
- Linkage mechanisms are primarily concerned to facilitate their work within these sub-systems and improve flows of information and technology with other sub-systems.
- Most linkage mechanisms have been operational and informal by nature. There are very few cases involving the creation of formal, permanent coordinating committees or of mechanisms giving NGOs and NARS representation in each other's directorate.
- Maintaining linkages requires resources.

A first clear implication of these patterns is that NGOs require, and have indeed sought out, more effective support from institutions operating in the applied and basic research sub-systems. One way or another, NGOs need NARS-type institutions and universities. A second implication is that if linkages are to be expanded in the future, this will have a resource cost, for both NGOs and NARS.

The patterns also suggest that, whilst there is now some experience of linkages on which to assess future possibilities, little has yet been done to create mechanisms that genuinely give NGOs voting power over decisions in the NARS. In this respect little has been done to democratize the NARS in any formal sense that gives rural voices, other than those of the larger farmers, a chance to hold them to account. There are cases of this happening in informal ways, and at an operational level, but not at the levels of policy formulation and institutional decisions on general resource allocation.

In the future, then, effort will have to be made to earmark resources for the maintenance of linkages, and above all to install structural linkage mechanisms. The changes occurring in the NARS described in Chapter 6 suggest that this is now possible, although it will not necessarily be straightforward.

Table 9.1 NGO–government relations without and with institutional contact

(a) **Without institutional contact**

Effect of NGO action on NARS work	*Nature of NGO activity*
Complementary	Fills gaps in ATS (people, place, thematic and sub-system gaps)
Critical	Criticizes government work in ATS, or government impact on context of ATS

(b) **With institutional contact**

Relation between NGO and NARS roles	*Linkage mechanisms*	*Comments*
Complementary: • between sub-systems • within sub-systems	*Operational*: • planning and review processes • collaborative activities • resource allocation procedures • communication devices • training activities	Can be formal or informal (usually informal)
	Structural: • coordinating units • permanent committees • representation	Must be formal
Critical	*Operational* (as above) *Structural* (as above) *Policy debate*	Where criticism and complementarity are combined, linkage mechanism is placed under increasing stress

Interaction for critique

The linkage mechanisms discussed in the preceding section have simultaneously been used by NGOs to express criticism of what government is doing within the ATS, and of policies that the NGOs perceive to be prejudicial to the rural poor. The contacts, mutual knowledge and fora that such mechanisms open to NGOs offer a chance to combine operational discussions with more fundamental debates. The same applies to NARS who are equally able to voice criticisms of NGOs through these linkage mechanisms.

The levels at which such critical observations can be made vary, with correspondingly different degrees of impact. If an NGO technician criticizes the technologies with which a local NARS field agronomist is working, the impact will go little further than the locality and will not address the structural and institutional factors limiting what can be achieved through the public sector. Critical debates in permanent committees at regional and national levels offer the possibility of a more widespread impact.

On the other hand, the evidence tends to suggest that local interactions have thus far been more successful than national ones based on structural linkage mechanisms. The success of structural mechanisms depends on the existence of favourable contextual factors, as are currently found in Chile, for instance where state and NGOs share much common ground regarding social development policy and where the relationship between the two has been close and collaborative. In other contexts the possibility of exploiting the potential benefits of structural linkages may be much more limited.

Which strategy to pursue, the local or the structural, is a matter of political choice. Those like Chambers favouring strategies of institutional and policy change based on 'small steps and little pushes' (Chambers 1983: 192) might prefer the more incremental localized approach to change; political economists may believe this will never make any impact on the underlying policies and structures that must be changed.

Either way, what is clear is that NGOs' role as (perhaps the most) highly informed critics of government must not be lost; rather, they must find more effective mechanisms of making their criticisms heard.

The problem of combining complementarity and critique

We noted that NGOs have tried to express criticisms of government programmes through linkage mechanisms that they are already using to enhance institutional complementarities and programme effectiveness. The problem with such a strategy is that it places these linkages under considerable stress, with negative implications for their effectiveness. It can even lead to the breakdown of such mechanisms. More generally, for an NGO to criticize the NARS at the same time as it is coordinating with it can induce tensions in the overall relationship.

Given this, it might sometimes be more effective to seek alternative mechanisms for expressing criticisms. Such mechanisms might be general policy discussion fora, such as those that the research NGO ILDIS (the Latin American Institute for Social Research) arranges in Bolivia. We have also found evidence of some NGO staff using informal channels and contacts with senior officials, politicians and donors as the vehicle they prefer for exerting influence over policy. Such channels give the NGO no formal sanction over the state – but it is our impression that they have been most effective.

In this respect, the experience gained during the period of this research project is informative. Most of the agencies involved used the preparation of the case studies and the Santa Cruz workshop to make criticisms of other institutions. The workshop in particular then seemed to help clarify certain misperceptions of the difficulties under which the different institutions worked, leading some to revise some of their criticisms. It also led to conversations that may lead to more complementary forms of interaction between these institutions. The creation of an informal and unofficial environment gave people the opportunity to explore ideas and express concerns that it would have been difficult to discuss in more operational relationships. Ironically the workshop also brought together in Bolivia people from NGOs, public-sector agencies and development foundations who had not talked to each other in their own countries. Opinions were undoubtedly softened.

An alternative strategy for combining critique and collaboration that was also noted in the Asian study of NGOs (Farrington and Lewis 1993) is for only certain NGOs to play the role of policy critics. This clearly happens. Research NGOs, who are barely or not at all engaged in operational activities, pursue this function.[3] However, for their critique to be informed and incisive, their links to other operational NGOs must be good; and for their critique to have an impact, they must have more powerful means of communicating it than a mere publication. Once again, informal contacts at a high level are crucial in this regard. The other possibility is that NGO networks assume this critical role. UNITAS-PROCADE in Bolivia is a good instance of this, and is quite closely involved in policy discussions with Bolivian government. However, the more that governments and donors push these networks into an operational and fund-disbursing role in which they channel money to their members, the more the networks' freedom to criticize may be compromised.

Once again, our conclusions are guarded, presenting pros and cons. What, however, does seem clear is that: (1) NGOs should protect this role as policy critic; (2) they can do it more effectively the closer their informal contacts with senior officials; (3) donors should continue supporting agrarian research NGOs, and should foster informal links between all involved if they want to sustain NGOs in this role.

Institutionalizing interaction

The issue of how far interaction can be institutionalized is moot. Most of the instances of linkages in the studies have been informal. The advantages of this are the flexibility it gives to individuals, making all interactions speedier than they would be if they were mediated by formal procedures. Similarly, informal links build on personal relationships of mutual trust, thus resolving many of the uncertainties institutions have *vis-á-vis* each other.

This dominance of informal interactions was emphasized repeatedly throughout the research, as were their benefits. Yet, at the workshop in Santa Cruz, the fragility and non-sustainability of informal arrangements were also stressed. As soon as an individual leaves the post in question, the arrangement comes to an end. There are also limits on what can be achieved as regards, for instance, resource allocation through informal arrangements.

The implication is that some degree of formalization, with its attendant costs of bureaucratization, is necessary if linkages are to be properly and fully exploited. The issue, then, is what type of formalization, when and where.

The research suggests that formalization of coordinating and other linkage mechanisms is likely to be more flexible, transparent and adapted at a local level. If the need to enter formal national arrangements is subsequently felt, it must be considered; but the evidence suggests the validity of starting locally. The current trend towards the decentralization of public administration can only favour this. The answer to the question of 'where?' will thus often be 'locally'.

'When' is more uncertain, but formalization will be most effective building on already established informal links whose participants begin to identify the need for more formal institutional arrangements. The answer to 'when?' is thus 'after a process of informal interactions'.

The question of what types of formalization cannot be answered *a priori*. Different mechanisms are appropriate for different circumstances and technologies (cf. Agudelo and Kaimowitz 1991). These will have to be identified locally, as and when they arise. Formalizing linkage mechanisms should thus occur mechanism by mechanism, as seems appropriate to the actors involved. The establishment of some sort of overall 'memorandum of understanding' (*convenio* in Latin America) may be all that needs to be done at a national or directorship level in order to encourage this more bottom-up (indeed participatory) approach to formalization.

The 'informal' is very current and fashionable in Latin America. None the less, building a new relationship between state and society that allows sustainability, accountability and mechanisms for demanding transparency necessarily requires institutionalized, formal procedures (Lehmann 1990).

IMPLICATIONS FOR DONORS AND ACTORS IN THE ATS

Implications for NARS and NGO managers

In the foregoing and Chapter 7 we have drawn several conclusions that have implications for NGOs and NARS managers. The overall message is that much can be gained from entering closer forms of inter-institutional relationship, and that the mechanisms for this currently in place are insufficient, inadequately specified and poorly targeted. On the other hand, establishing these mechanisms has a resource cost, and therefore care must be taken to enter only those that are most appropriate, and then to budget for them accordingly.

Such mechanisms cannot, however, be pulled out of the magician's hat (or some multilateral donor agencies' Cookbook for Success). They have to be worked at, and much attention has to be paid to generating a favourable context. Perhaps the most central element in this context is a believable policy commitment to the resource-poor farmer: NARS have to show NGOs that they have genuinely made this commitment, and NGOs must show NARS that they are more concerned about this than they are about ideological purity, party political or their own institutional concerns. Second is the need for mutual understanding and trust. This can only be built on more frequent, honest and open contacts. Before proposing mechanisms, managers and staff must therefore spend much effort visiting each other's field work and offices. Just as farmer participation and 'undistorted' communication must be based on the stripping away of distrust, misperceptions and power relations, so must inter-institutional communication. Third, interactions and linkages will be more effective, and more possible, when a general culture of collaboration dominates in the institutions involved. Building this culture is a task that falls on managers.

Entering such inter-institutional relationships implies quite important changes for NGOs and NARS alike. These will at times be painful, and will certainly require soul searching regarding institutional identity and roles, as well as reflection on the development model and social contract the institutions believe in. This soul searching and reflection cannot, however, be avoided. It is part of a more general change in which roles and models are in question. All development institutions ignore examining such questions at their peril. A series of other implications for NARS are also particularly relevant for donor agencies and they follow in the next section.

Implications for donors

Given that certain donors are beginning to push these new inter-institutional roles, they have a particular responsibility to think carefully about their implications. The following seem particularly pertinent points on which to reflect.

From instrumentalism to learning

We have referred on several occasions to the tendency of multilateral and many bilateral donors to see NGOs as a useful resource for doing jobs previously done by the state, either with their own resources or on some sort of contract to the state (and essentially to the donor providing the budget line). Aside from the concern as to whether or not NGOs will accept such a role, the point is that this seems a mistaken approach to the issue. NGOs ought to be a source of lessons on how to reconstruct a crumbling public sector; they ought not to be 'putty' to stuff the cracks.

NGOs can offer many lessons, but perhaps above all as regards institutional structure. Their internal organization and relations to the outside world, while imperfect, have been assets for their flexibility. This institutional structure has in turn been important in allowing them to use and adopt certain work methods, and certain lower-input, food security-based and farming-systems-based technologies that are central ingredients to the 'new technological agenda' we talked of in Chapter 4.

By looking carefully at NGOs' prior experiences in all these areas, governments and donors could learn a great deal as they think through the restructuring of a decentralized public sector, and a strategy of agricultural development that addresses sustainability and poverty concerns.

From institutional fixes to policy commitments

The importance of restructuring rather than patching up is relevant to a second point: NGOs ought not to be seen as institutional fixes to a problem of policy. Policies require appropriate institutions to implement them, but the policy commitment to the rural poor must come first. Again, this is not only because NGO attitudes to coordination with government will depend on this policy commitment; it is also that without it, NGOs' work will too often have limited life and impact.

Thus, we have suggested that NGOs do need the support of a NARS or some other institution that provides the same services and makes them economically accessible to NGOs. Many NARS in South and Central America are in a critical condition: a policy commitment to strengthening their resources and ability to serve the rural poor is essential if NGOs are to be able to do an effective job. It is not an either/or question: NGOs or NARS. There are many other similar examples that we could give, and they lead us to the following general points:

- NGOs are in no way solutions to underdevelopment;
- support to NGOs should not be used as a smokescreen to hide the fact that public institutions such as NARS are being run down;
- the effectiveness of linkages with NGOs will be enhanced for both sides when public institutions are professionally and technologically strong.

A second point for donors is that the whole legitimacy of the NGO label is challenged by the proliferation of opportunistic institutions calling themselves NGOs, but whose hearts often seem closer to their bank accounts than to the rural poor. Much of this proliferation is a direct effect of donor actions which create unemployment among NARS and public-sector professionals, and which then provide money for NGOs that are spontaneously formed to take advantage of that money.[4] Alongside the new-found enthusiasm for NGOs, one can already sense a rising 'malcontent' with them as a result of this proliferation. NGOs are in danger of being drowned in too much donor love.

From devising blueprints to recognizing diversity

Finally, it cannot be stressed enough that the non-government sector is an extremely heterogeneous community, which has the potential of performing many different roles. The new opportunist NGOs may be willing to be contractees of the state, older radical NGOs may be more circumspect; research NGOs may be better placed providing constructively critical advice on the ATS, and operational ones more suited to coordination with NARS at an implementational level; NGOs whose staff genuinely suffered at the hands of the state may be more wary of new relationships than those that have not.

In many respects, this diversity could be an asset to the ATS, but this potential will not be exploited by squeezing NGOs into blueprinted roles such as service delivery. Inflexible and clumsy interventions not only fail to take advantage of this diversity – they can also damage it. For example, channelling plentiful donor money through one network and not another, or only through networks rather than non-network members, runs the risk of marginalizing those NGOs who do not enjoy such generosity. As those who benefit grow and pay higher wages, so those NGOs on the 'outside' will lose staff to these higher-paying employers, unravelling one institutional resource and creating another – with valuable energy (and knowledge) expended in the process.

Planning for diversity is of course more complex than drawing up and multiplying blueprints. It implies a process conducted closer to field and institutional realities. As donors press governments to decentralize in order to be more adaptive and flexible, they might themselves be well advised to heed their own advice.

A final implication for all actors in the ATS

Finally, we would reiterate that different linkage mechanisms and interactions serve different ends. NGOs, NARS and donors alike have to decide exactly what it is they want of each other, and exactly the roles they feel each other ought to perform. To decide this they have to be clear on their approaches to and models of development, and on their strategies as institutions within these models. Too many organizations are too unclear on these models. Much hard thinking on 'bigger' questions must come first if inter-institutional relations are to be best exploited for reasons of strategy rather than convenience.

NOTES

1 Echenique and Rolando (1989) book, reflecting AGRARIA's analysis of the peasantry in Chile, uses the terminology of a 'Social Debt' in its title.
2 One of the most striking features of the workshop in Santa Cruz was the willingness of all involved to admit their failings – and particularly the willingness of many NGOs to recognize that their performance fell short of their rhetoric.
3 For instance, CEDLA in Bolivia, GIA in Chile, CEPLAES in Ecuador and Celater in Colombia.
4 In the workshop, one working group talked of NGOs 'sprouting like mushrooms'.

ANNEX 1

CASE STUDIES COMMISSIONED DURING THE RESEARCH

The research in this book was based largely on case studies commissioned during the course of the study. These are listed below. Following the title of the paper, the institution to whose experiences it refers is given in parentheses.

- Aguirre, F. and Namdar, M. 1991. 'Relaciónes Sector Público-ONG. El Caso de AGRARIA' (AGRARIA-Chile; Providencia No. 1387, Piso 3, Santiago, Chile. Tel: 493530; Fax: 562 2352459, Telex: 240301 BOOTH CL).
- Altieri, M. and Yurjevic, A. 1991. 'Influencias de las Relaciónes Norte-Sur en la Investigación Agroecológica y Transferencia de Tecnología en América Latina: El Caso de CLADES' (Consorcio Latinoamericano de Agroecología y Desarrollo, CLADES-Latin America; Casilla 16557, Santiago, Chile. Tel: 2341141; Fax: 2338918).
- Bojanic, A. 1991. 'La Transferencia de Tecnología en Bolivia: La Marcha para llegar al Modelo de Usuarios Intermediarios' (Instituto Boliviano de Tecnología Agropecuaria, IBTA–Bolivia; Casilla 5783, Plaza España Esq. Mendez Arcos, No. 710, La Paz, Bolivia. Tel: 326996; Fax: 370883).
- CAAP, 1991. 'Generación y Transferencia de Tecnología Agropecuaria. Sistematización de Experiencias en el CAAP' (Centro Andino de Acción Popular, CAAP – Ecuador; Diego Martín de Utreras 733 y las Casas, Apartado Postal 173B, Quito, Ecuador. Tel: 522 763).
- Cardoso, V.H. 1991. 'Relación del Programa de Investigación en Producción del INIAP con las Organizaciónes No Gubernamentales y Organizaciónes Campesinas' (Instituto Nacional de Investigaciónes Agropecuarias/Programa de Investigación en Producción, INIAP-PIP–Ecuador; Casilla 2600, Av. Eloy Alfaro y Amazonas, Ed. MAG 4to.P., Quito, Ecuador. Tel: 565939/565963).
- CESA, 1991. 'La Relación de CESA con el Estado en la Generación y Transferencia de la Tecnología Agropecuaria' (Central Ecuatoriana de Servicios Agricolas, CESA–Ecuador; Casilla 16-0179CEQ, Calle Inglaterra No. 532, Quito, Ecuador. Tel: 524-830 546-606).
- Chavez, J. 1991. 'Los Programas de Agricultura Ecológica y de Agro-

Industria Alimentaria del Centro IDEAS' (Centro IDEAS–Perú; Casilla 11-0170, Avenida Arenales 651, Lima 11, Perú. Tel: 2477737).

- Estrada, J.F. 1991. 'Análisis Evolutivo en el Proceso de Transferencia de Tecnología en el Instituto Colombiano Agropecuario ICA: Una Experiencia Institucional' (Instituto Colombiano Agropecuario, ICA–Colombia; Regiónal S, Apartado Aereo 10140, Calle 5A No. 38A-58 Imbanaco, Cali, Colombia. Tel: 570484).
- FUNDAEC, 1991. 'FUNDAEC y el Sector Público: Un Enfoque de Integración entre las ONG y Organizaciones Gubernamentales para la Generación y Transferencia de Tecnologías Agropecuarias' (Fundación para la Applicación y Enseñanza de las Ciencias, FUNDAEC–Colombia; Apartado Aereo 6555, Carrera 41 5C-116, Cali, Colombia. Tel: 536489; Fax: 536491).
- García, R., Warmenbol, K and Matsuzaki, S. 1991. 'Transferencia de Tecnología y Desarrollo Comunitario: Experiencia de CIPCA en el Departamento de Santa Cruz, Bolivia' (Centro de Investigación y Promoción del Campesinado, CIPCA–Bolivia; Casilla 3522, Av. 26 de Febrero 652, Santa Cruz, Bolivia. Tel: 347366).
- Gonzalez, W. 1991. 'PROCADE: Un Análisis de Su Rol Coordinador de Actividades Interinstitucionales y de Relación con el Sector Público Agropecuario' (Programa Campesina Alternativa de Desarrollo, PRO-CADE–Bolivia; UNITAS, Casilla 8666, Calle Abdón Saavedra No. 2323, La Paz, Bolivia. Tel: 353048).
- Gonzalez, W. 1991. 'Diferenciación Institucional en el Manejo de Tecnologías Agrícolas en la Región del PROCADE' (Programa Campesina Alternativa de Desarrollo, PROCADE–Bolivia; UNITAS, Casilla 8666, Calle Abdón, Saavedra No. 2323, La Paz, Boliva. Tel: 353048).
- Gonzalez, R. 1991. 'Generación y Transferencia de Tecnología Agropecuaria: el Papel de las ONG y el Sector Público: El Caso del Secretariado Diocesano de Pastoral Social 'SEPAS' San Gil, Santander, Colombia' (Secretariado Diocesano de Pastoral Social 'SEPAS'–Colombia; Apartado Aereo 44 San Gil, Carrera 9 No.13-07 San Gil, Santander, Colombia. Tel: 97724 2295/2573; Fax: 97724 3393).
- Guerrero, L. 1991. 'La Generación y Transferencia de Tecnología entre ONG-Estado-Universidad en el Departamento de Cajamarca' (Centro de Investigación, Educación y Desarrollo, CIED–Perú; Las Casuarinas F3, Urb El Ingenio, Casilla 131, Cajamarca, Perú. Tel: 922924; Fax: 923429).
- Kaimowitz, D. 1991. 'El Papel de las ONG en el Sistema Latinoamericano de Generación y Transferencia de Tecnología Agropecuaria' (D. Kaimowitz, Programa de Generación y Transferencia de Tecnología, Instituto Inter-americano de Cooperación para la Agriculture, IICA-Costa Rica; Apartado 55-2200, Coronado, Costa Rica. Tel: 290222).
- Kohl, B. 1991. 'Sistemas de Cultivos Protegidos en el Altiplano Boliviano:

Proyectos de Tecnología Apropriada en las ONG' (B. Kohl, Casilla 5865, Cochabamba, Bolivia).

- Kopp, A and Domingo, T. 1991. 'Tecnologías de Conservación en el Trópico: CESA, Bolivia' (Central de Servicios Agropecuarios-Bolivia; Asesores en Asuntos Campesinos, Casilla 4691, C. Colombia 275, Plaza San Pedro, La Paz, Bolivia. Tel: 343233).
- Sotomayor, O. 1991. 'El Estado y las ONG Chilenas' (Grupo de Investigaciónes AGRARIAS-Chile; Casilla, 332-Correo 22 Santiago, Ricardo Matte Perez No. 0459, Santiago, Chile. Tel: 2255636/2047432; Fax: 2235249; Telex: 343351 GIA-CK).
- Torres, R. 1991. 'Organizaciones Nongubernamentales y Sector Público en el Perú: La Experiencia de la Comisión Coordinadora de Tecnología Andina' (Comisión Coordinadora de Tecnología Andina, CCTA-Perú; Av. Javier Prado No. 595, Lima 17, Perú. Tel: 617253; Fax: 014 421766).
- Trujillo, G. 1991. 'Investigación y Extensión por El Ceibo en el Alto Beni.' (Central Regiónal Agropecuaria-Industrial de Cooperativas 'El Ceibo' Limitada, El Ceibo-Bolivia; Casilla 9698, Avenida Juan Pablo II No. 2560, La Paz, Bolivia. Tel: 812817).
- Veléz, R. and Thiele, G., 1991. 'Primeras Experiencias con un Nuevo Modelo de Transferencia de Tecnología' (Centro de Investigación Agricola Tropical, CIAT-Bolivia; Casilla 247, Avd. Ejército Nacional 131, Santa Cruz, Bolivia. Tel: 342996).
- Zeller, T. 1991 'COTESU y las ONG.' (Cooperación técnica Suiza/Swiss Technical Cooperation, Bolivia; Casilla 4679, Calle R. Gutierrez No. 4679, La Paz, Bolivia. Tel: 340168).

In the United Kingdom the papers are available at the Overseas Development Institute, ODI, Regents College, Regents Park, London, NW1 4NS. In Latin America copies are held at the Centro de Investigación Agricola Tropical/ British Tropical Agricultural Mission, Casilla 247, Santa Cruz, Bolivia, and the Centro Latinoaméricano de Tecnología y Educación Rural, Apartado Aereo 020756, Cali, Colombia.

Several of the papers have been published in English as Network Papers by the Agricultural Research and Extension Network at ODI. They are:

Aguirre, F. and Namdar, M. (1992) 'Complementarities and Tensions in NGO-State Relations in Agricultural Development: The Trajectory of AGRARIA, Chile', *Agricultural Research and Extension Network Paper 32*, Overseas Development Institute, London.

Kohl, B. (1991) 'Protected Horticultural Systems in the Bolivian Andes: A Case Study of NGOs and Inappropriate Technology', *Agricultural Research and Extension Network Paper 29*, Overseas Development Institute, London.

Sotomayor, O. (1991) 'GIA and the New Chilean Public Sector: the Dilemmas of Successful NGO Influence over the State', *Agricultural Research and Extension Network Paper 30*, Overseas Development Institute, London.

ANNEX 2

CASE-STUDY PROFILES

While we have incorporated much of the case-study material into the text, it seemed helpful to provide several somewhat more comprehensive study profiles of NGOs' work in ATD as synthesized by them in the background papers they prepared for this research. In this annex we therefore present some illustrative case-study profiles. These are selected not because we deem them any more important than the others but because they indicate both the range of, and similarities between, the approaches that different NGOs have taken to agricultural technology. The profiles are based on summaries of papers the NGOs themselves prepared, along with some information generated from interviews with their staff. In that regard, they primarily represent 'their' view rather than our own.

The first three profiles are of NGOs we denominated in Chapter 4 as modernist. Profiles 4, 5, 6 and 7 are of NGOs working with a more agroecological perspective, and the last profile is of an 'ethnodevelopment' NGO.

PROFILE 1 AGRARIA

Chile

AGRARIA: Desarrollo Campesino y Alimentario
(AGRARIA: Food and Campesino Development)
by
F. Aguirre and M. Namdar-Irani

AGRARIA and campesino development

AGRARIA was established in 1983 as a non-profit professional association with the main objective of supporting small-scale producers. It implements research and development programmes in micro-regions containing significant concentrations of campesinos. These projects have developed four main types of activity: agronomic and socioeconomic research, technical support, credit and community organization.

More recently, AGRARIA has undertaken what it calls 'second level' activities. These aim to introduce more fundamental ('structural') changes into production systems such as irrigation and afforestation. They also seek to improve the terms of campesino integration into the market through processing and marketing activities. AGRARIA's extension work is now mainly financed by contracts with the government's Institute of Agricultural Development (INDAP). Its 'second level' activities are funded by foreign donations and national development funds.

Currently, AGRARIA has a team of around 100 professionals and technicians and works with 3,400 campesino families, of whom 2,200 participate in the INDAP extension programme.

Parallel to these activities, AGRARIA is carrying out an analysis of its own experience. It is also evaluating nutrition problems and broad trends in Chilean agriculture, with the goal of stimulating debate on Chilean agrarian policy. Lastly, it has recently expanded its training activities beyond its own staff and now gives courses to professionals from other institutions in the Chilean public sector and from other Latin American countries.

The history of one project in the micro-region of Sauzal, located in the 7th Region of Chile, demonstrates some of AGRARIA's more general experiences in technology development and its relationships with government. It shows that in its early years the NGO made many of the mistakes frequently criticized in government programmes: it worked with high-cost, yield-maximizing, commodity-based technologies that were inappropriate to farmers' needs. In response to these errors, AGRARIA began to operate with a more systems-based approach to small-farm development, incorporating both on- and off-farm concerns into its interventions, and combining technical support with other activities such as training, credit and organizational development. As AGRARIA began to work towards introducing structural as well as functional changes into farming systems, it felt the need to collaborate with other institutions for research inputs and other forms of support.

Until 1990, formal links with state institutions were largely precluded by the political context, but informal relationships and research collaborations supported AGRARIA in its elaboration of new technical proposals for farmers. When Chile returned to electoral democracy in 1990, relationships with government improved dramatically, but they have generated new challenges for AGRARIA, which are discussed below.

The Sauzal Project

The micro-region of Sauzal is located in the dry interior of Cauquenes. It covers some 3,500 square kilometres, with a predominantly rural, thinly dispersed population of around 64,000.

The Sauzal micro-region is a pasture zone where extensive sheep-farming predominates. All work is carried out with animal traction. Large and medium farms (of over 50 hectares), some 800 properties, represent only 10 per cent of landholdings but cover 75 per cent of the land. These farms use hired labour and/or share-cropping, and their principal products are cattle, sheep, vines (mostly in the flat-lands), wheat and pines. In the last ten years, a significant number of these farms have been bought by forestry firms and have been completely forested. Small and medium family farms comprise some 7,000 properties and cover 25 per cent of the land. Their activities include wheat cultivation on the hills (for domestic consumption and sale); pulses (especially chick peas produced for consumption in the hilly micro-region, and for the market in the fertile lowlands); vine cultivation; sheep and goat farming, and sometimes cattle-farming. The campesino farming sector is not homogeneous, with the quantity and quality of land managed varying widely.

This zone has always been excluded from the government development initiatives. It has few roads and irrigation works, and its productive and social infrastructure is deficient. The consequence has been out-migration which has led to a stable, and in some sub-zones a declining, population since the turn of the century.

Main stages in the project's evolution

The project has passed through four main stages as it developed:

Stage 1: familiarization with the zone (1984–5). This stage involved socioeconomic and agronomic evaluations of campesino production systems. These first appraisals were too brief and insufficiently detailed, leading to subsequent errors in research and technical support.

The first stage emphasized strengthening campesino organization. Seven geographically based committees were set up, grouping 150 campesino families. This organizational work was complemented by support activities such as vaccination campaigns, and some demonstrations of alternative cultivation techniques in crops sown by a majority of the campesinos, principally wheat.

Innovations in this first stage were characterized by their simplicity and their short-term impact. They were linked to a small credit operation handled by DAR, the Rural Action Department of the Bishopric of Linares, which had worked in the area since 1978.

During the first two stages, AGRARIA also worked on developing communication and education methods with help from an NGO specializing in popular education.

Stage 2: improving campesino food self-sufficiency (1986–7). In this stage, research was extended into wheat and into the diversification of campesino food production through the introduction and improvement of pulse cultivation (lentils and chick peas) and small-scale vegetable production. This agronomic research served as the basis for the technical assistance support work carried out with groups of campesinos.

These activities were strengthened by improvements in the sub-system of animal traction, through work on animal health (vaccination campaigns twice a year and the training of veterinary paratechnicians), the search for improvements in animal feed and the introduction of forage crops (forage oats and sorghum).

The attempts to strengthen campesino organizations continued with the voting in of committee executives and an increase in the number of families involved (230 campesino families organized in eleven committees).

Substantial changes which characterized this stage were:

- a shift from demonstration activities to adaptive research (with a minor excursion into basic experimental research);
- a diversification of the types of activity addressed in the research. This involved research into new sub-systems, and improvements in sub-systems interrelated with a number of others, such as animal traction.

Stage 3: developing production systems as a whole (1988–90). From the start of 1988, support was differentiated according to the different types of production system. Two sub-systems were prioritized: vine cultivation and sheep-farming. These sub-systems were found in an important number of campesino production systems and were the basis for the potential development of small-scale agriculture in the zone. Agronomic research focused on adapting techniques elaborated by the national agricultural research institute INIA.

Since these activities did not interest all producers, 'interest groups' were established. These 'interest groups' were self-selected groups of producers with a special interest in a particular part of the production system.

This stage required greater credit support than the previous stages, as it dealt with perennial crops requiring initial capital investments for start-up costs. Therefore AGRARIA created a small credit-investment fund, which it continues to administer.

The number of beneficiaries grew steadily during this period and reached a total of 450 campesino families. This led to the decision to engage in agricultural development at a micro-regional level. This proposal included several lines of action:

- research and technical support differentiated according to agro-socioeconomic conditions;
- basic agronomic research;
- the promotion of medium-term technological improvements;
- credit for the promotion of small productive investments.

Stage 4: improving links between small producers and the wider socioeconomic environment. This stage has been facilitated to a large extent by the election of a new government and its decision to devote more resources to small-scale agriculture. As AGRARIA has been able to capture state resources to fund technical-assistance work, it has begun to place more emphasis on improving the terms on which campesinos are integrated into the market. It has therefore initiated a wheat marketing programme in which AGRARIA helps campesinos sell their product to a national parastatal marketing agency, and then helps them recover value-added tax. A system providing agricultural price information to give small producers more information before they negotiate the price of their products is also planned as part of this institutional intervention in spheres of marketing. Also, it is hoped to introduce processing technology for some products (e.g. chick peas) with the aim of increasing the value added during the stages of production controlled by the campesinos.

Another new goal of the project has been to help small producers to gain improved access to forms of state support that the new government has now made available to campesinos. This support is both productive (irrigation and forestation) and social (guarantee of title, housing, pension). AGRARIA provides information and represents campesino interests *vis-à-vis* state institutions.

Assessment and perspectives

After seven years' work, an assessment of the main gains and limitations of the project can be made.

For AGRARIA as an institution the main gains have been:

- strengthening and enrichment of its methodology for work with small producers;
- the successful integration of the institution in the micro-region;
- success in gaining recognition for local producers' organizations by other national institutions.

For producers in Sauzal the main gains have been:

- improvement in techniques of wheat production;
- improvement in animal health;
- proposal and pilot project activities related to vine-cultivation, livestock and poultry farming;
- elaboration of proposals for micro-regional development.

Changes since the return to democracy in 1990

Gains such as these would have levelled off earlier had there not been the radical changes in the national political situation of 1990. The activities of AGRARIA during the years of dictatorship were focused on production improvements which did not require large investments. Its work did not extend to 'second level' activities such as marketing and processing. Similarly, problems identified by AGRARIA which affected rural people beyond the immediate sphere of agricultural production could not, at that time, have been communicated to public-sector institutions.

Since 1990, AGRARIA has been selected by the government as an agency for INDAP's technology-transfer programme in various zones of the country. In the dry-land interior of Cauquenes, it won contracts in 1990 and 1991 to deliver technical support to 430 producers. In addition, demonstrations and trials have been carried out. These teams are financed by INDAP. Their work programmes are based directly on the cumulative experience of AGRARIA from previous years.

AGRARIA experienced substantial growth from 1990 onwards as a result of its participation in the INDAP extension programme. The programme is important to AGRARIA for several reasons:

- It provides a degree of financial stability and enhances AGRARIA's status.
- It allows AGRARIA to expand into many new areas, thus broadening its experience of campesino conditions nationwide.
- It offers the possibility of freeing up funds with which to develop new 'second level' activities.

However, the programme has also generated a number of problems. INDAP's approach has methodological errors and weaknesses and has been too rigid to change in response to field experience. Therefore it is imperative that a forum be created in which AGRARIA, along with other institutions who have experience in this field, can discuss methods and strategies with the relevant public bodies.

Another problem is that the rapid growth of AGRARIA led it to hire personnel who do not necessarily have the appropriate methodological basis or experience. AGRARIA must pay more attention to training these new staff if full advantage is to be taken of the approaches and methods it has developed during its many years' experience.

An institutional reflection on public-sector–NGO relations

The government's increased interest in the rural sector – particularly small-scale agriculture – has resulted in an increased availability of public-sector human and financial resources for rural development. The increased number of institutions involved in rural development and their new orientation towards campesino agriculture imply an increased need for coordination to ensure coherent and balanced development.

To finance the new activities the state has set up a Cooperation Agency to secure funds to which until recently only NGOs had access. Furthermore, the election of a democratic government has produced two important changes in international funding policy: more resources are available, thanks to an increase in bilateral cooperation, and the Chilean government now enjoys favoured access to these resources. As a result, competition for resources between NGOs and the state has increased, in large part as a product of the poor definition of how roles and responsibilities should be divided between NGOs and the public sector.

So far, the role of NGOs under the new government has largely been limited to implementation of public programmes, using their existing organization and experience, but not allowing them to participate to any significant extent in debates over policy definition.

PROFILE 2 CESA

Ecuador

Central Ecuatoriana de Servicios Agricolas
(Ecuadorian Centre for Agricultural Services)
by
CESA

The origins of CESA

In Ecuador, the 1960s saw the development of numerous campesino organizations pushing for agrarian reform and working in contact with the trade-union movement and left-wing groups. It was in this context that CESA was founded in 1967 by the leadership of several class-based organizations, initially to undertake a pilot project in agrarian reform on land owned by the Catholic Church. At the outset its main aim was to promote campesino organization in order to ease the integration of the rural poor into existing models of development. Over time, however, CESA's focus shifted slightly from simply promoting market integration to strengthening the capacity of campesinos to negotiate relations with both the market and the state. A key part of this work has been to promote campesino organization via consciousness raising and educational activities. This has been combined with more specific activities introducing new production technologies along with methods for product processing and input supply which are intended to give campesinos greater control over agricultural production and marketing. The work is thus at once economic and political.

With this set of targets and ideas, CESA is currently engaged in the following areas of work: training in community organization, ATD, forestation and natural-resource conservation, infrastructural provision, and marketing.

ANNEX 2

CESA's perspective on ATD

CESA takes an intermediate position on the debate over the roles of traditional and modern technology in Ecuador. At one extreme, CAAP, the Andean Centre for Popular Action (see Profile 8) argues for a technological alternative based on traditional agroecological practices. At the other extreme, INIAP (the public-sector NARS) has long worked with Green Revolution technologies. CESA prefers to start from campesino practice, which, in general has already incorporated agrochemical technologies and 'modern' varieties.

CESA has a specific programme of agricultural experimentation and demonstration which, in seeking to reduce the intensity of external-input use, is essentially working to adapt INIAP recommendations to campesinos' circumstances. The programme operates in CESA's ten different project areas in Ecuador, covering in all some 25,000 hectares and a total of 150 grassroots campesino organizations (more than 8,000 families).

ATD in CESA and contacts with the public sector

CESA sees its experiment/demonstration programme as filling a gap left in agricultural technology development and transfer by the state. In CESA's opinion, public-sector institutions are doing little or no work to adapt their recommendations to campesino conditions, and their programmes have failed to offer consistent support to campesinos. CESA's aim is not just to fill this gap, but also to build links with public and private institutions and exercise some influence over their technological agenda.

In this strategy of building links, CESA has two objectives: to gain access to the state's scientific and technological resources, and to harness the considerable institutional potential of INIAP and other public institutions for the benefit of campesinos. Often CESA's links with public-sector bodies have remained informal, as in its links with the on-farm PIP research programme, but they have sometimes been formalized, and between 1976 and 1986 CESA had several specific formal agreements with INIAP. For example, there were agreements for INIAP to train CESA paratechnicians (1976), and for the joint establishment of on-farm demonstration and experimental plots.

This strategy of reaching a separate agreement for each individual collaboration proved excessively time-consuming, so to streamline this relationship, in 1986 CESA requested a general agreement with INIAP. This allowed greater flexibility at local and operational level, although in and of itself it was insufficient to create collaborations: it merely facilitated initiatives.

The quality of these CESA–INIAP relations has been uneven, due to the internal institutional problems of INIAP. In the mid-1980s, INIAP entered a period of crisis and institutional change as a result of efforts to privatize it. Other obstacles to establishing links are that INIAP suffers from high staff turnover; that it focuses primarily on developing modern technologies and export crops; that it moves very slowly and is reluctant to exchange information.

The most successful CESA–INIAP partnership to date has been in the generation of fruit-growing technology for small farmers. The leverage role of a donor was an important factor in this success. The Swiss bilateral aid agency COTESU supported an INIAP project of research on fruit trees. Simultaneously, COTESU was supporting a campesino development project with CESA in Chingazo-Pungal (Chimborazo province) where there was potential for enhancing campesino income by improving the technology for growing fruit. However, CESA was unable to build on this potential because it lacked the skilled personnel necessary and had no prior experience in fruit growing. To resolve this bottleneck, COTESU took advantage of its ties with INIAP and tried to foster closer contact between the two institutions. This gave CESA the access it required to specialist research and technical assistance resources, and at the same time gave CESA the chance to propose modifications to the high-cost package that INIAP had been working with up to then. The result was that the research and technology transfer was adapted to campesino circumstances, and was subsequently adapted in the light of experience to incorporate slight innovations after discussions with campesinos. INIAP also provided a part-time expert to the CESA project, although lack of funds at INIAP created some problems and impeded the frequency with which he visited. Sometimes to speed these visits CESA would cover costs. Overall, the local impact of this project has been substantial and CESA is about to transfer tree nursery management to the campesinos.

There is also a more indirect link between CESA, INIAP and other public institutions. In its applied research CESA tests with campesinos the research results that have been generated in the research at INIAP (and other institutions). Often university staff and graduates also collaborate with CESA and participate in its projects as researchers. In one such link with the universities, graduates do research and write their theses on topics suggested by CESA. CESA provides logistical support. The partnership between CESA and the National Rice Programme (PNA) in Daule is a similar case. The PNA supplied an expert to join the CESA team for several years to give technical assistance to CESA paratechnicians and campesinos. CESA covered the operating costs and informed the PNA of campesinos' needs.

Dissemination of practices verified in these various projects for the adaptation and testing of technologies is undertaken with and by the campesino organization alongside which CESA works. Demonstration plots are established in communal lands, and pilot trials for seed production and multiplication are planted in communal lands. The maintenance of the trial and the distribution of the seed multiplied is coordinated by the federations uniting the communities in CESA's project area. The aim is to keep members supplied with seeds produced from the plot and to strengthen the federation in the process. CESA provides a link with INIAP, buying its improved varieties for multiplication in the plots – although on several occasions INIAP has been unable to provide CESA with the seed it has needed.

CESA has held shares in two parastatal companies, the Compañía Nacional de Seguros Agropecuarios (National Company for Agricultural Insurance) (CONASA), and the Empresa Mixta de Semillas (Joint Seed Company). Being a shareholder gives some scope for influencing company activities; at the very least it facilitates knowledge of the companies' activities and access to its resources. As one example, CESA used CONASA's services to provide insurance for the seed multiplication plots installed with the campesino federations.

CESA's greatest limitations in the experiment/demonstration programme have been shortcomings in planning, design, monitoring and follow-up of activities. It has also failed to systematize and quantify data on the management and impact of its programmes. These weaknesses resulted from methodological shortcomings and lack of computer equipment on which to build the requisite databases. Consequently, with advice from donors, a 'Planning, Monitoring and Evaluation Unit' has been created in CESA and a computer system installed.

CESA has also acknowledged that to improve the quality of its work and management systems, it needs to train its staff in technical and administrative skills and to modernize its administration.

Exerting influence through linkages with the public sector

While CESA's short-term aim has been to obtain access to public-sector technologies and then adapt them to campesino conditions, the longer-term aim has been to influence public-sector action, objectives and organization. CESA has to date had far less success in this more far-reaching objective.

CESA has pursued two strategies to achieve this goal. The first was to use its influence directly. For example, when INIAP set up an on-farm research programme (the PIPs), CESA not only saw the PIPs as a potential source of support for its own work, but also viewed them as an opportunity to influence INIAP's work. The idea was that CESA might be able to exert at a local level informal influence on the PIP paratechnicians, who would then carry those ideas back to the experiment station. However, the PIP field staff had little or no leverage over the work of the experimental station, and therefore CESA's potential influence was never realized.

A more positive experience stemmed from CESA's relationship with the Ecuadorian Agrarian Bank (Banco Nacional de Fomento). CESA established a joint fund with the BNF, under which CESA had responsibility for the management of small loans to campesinos who typically did not have access to BNF credits because of the onerous guarantee requirements. In this mode of operation, the campesino repays the loan to a CESA staff member who works one day a week at the BNF at a special CESA desk. The aim is not simply to distribute credit but also to introduce the campesino to the BNF (hence the insistence that repayments are made to CESA at a desk in the Bank), and more importantly to show the BNF that it is both viable and profitable to manage a small-loan scheme for campesinos. Although the programme has not been institutionalized by the BNF, it should be noted that the experience has been an important precedent to the design of a campesino credit system in the National Rural Development Programme (PRONADER), which commenced in 1992.

The second strategy of exercising influence is less direct. CESA supports campesino organization in the belief that once up and running, the organization itself will lobby for greater influence over public-sector actions and accountability in its decision-making.

Clearly the two strategies can be combined, as seen in CESA's role in an integrated rural development project in Canar. CESA had previously worked in the region strengthening the campesino organization, Tucayta. At the same time, once again thanks to COTESU (one of the project donors), CESA was represented and had voting rights in an inter-institutional committee which managed and monitored the project. Along with CESA, this committee included various public, sectional, regional and national institutions. CESA used its position on the committee to demand that Tucayta also be given representation and voting rights in the committee.

Today Tucayta is indeed a committee member, acting independently of CESA, and lobbies both NGOs and public institutions. In this case, then, CESA succeeded in achieving a significant institutional change, by creating a place for the campesino organization within the management of a public-sector project. More broadly, CESA hopes from this experience to show the public sector that it is viable and important to include campesino organizations at the level of management of activities (and not just at the stages of consultation and project implementation). CESA has pursued a similar strategy with the National Institute for Water Resources (INERHI), promoting the involvement of campesino organizations in irrigation systems management, the installations being co-financed by CESA and the public sector.

The aim is for the public sector to use this approach to involve campesino organizations in other projects and actions. To a certain extent this has been achieved in the design of PRONADER, where the campesino organizations will have a say in the management of the various PRONADER projects. Although CESA's direct influence upon the design of PRONADER cannot be proven, CESA did play an important role during consultations on project design. Furthermore, one of the general coordinators in the design of PRONADER had previously worked for CESA.

Overall assessment and constraints on CESA–state collaboration

Shared planning of activities between CESA and the public sector would be virtually impossible, given their diverse objectives, beneficiaries and style of working. However, in specific actions with concrete objectives it has been possible to agree on joint planning and implementation of specific activities. Here, it has been an institutional policy of CESA to seek links with the public sector to improve the technical support received by the campesinos. In general, CESA considers that, given the frequent overlapping of different institutions' activities, inter-institutional coordination is of overriding importance. At the same time CESA acknowledges that in working more closely with public institutions, there is an increased danger of being manipulated.

PROFILE 3 CIPCA

Bolivia

Centro de Investigación y Promoción del Campesinado
(Centre for Peasant Research and Promotion)
by
R Garcia, K. Warmenbol and S. Matsuzaki

This profile describes how CIPCA has changed its technology testing and transfer methodology as a result of its experience with campesinos. Simultaneously, it has redefined its role as regards technology transfer. CIPCA began carrying out its own adaptive research trials but concluded that it should limit its activities to the testing and transfer of technology which had already been developed in government research centres.

CIPCA began working in Bolivia in 1971. Its general aim is to assist campesinos in achieving full and active participation in Bolivian economy, politics and society. CIPCA works in two regions of the Bolivian highlands and two regions of the tropical plains of Santa Cruz: in the Andean Cordillera with Guarani Indians, and the migrant colonization area of the highlands. This profile discusses its work in the lowlands.

The CIPCA development model

CIPCA aims to develop campesino organization at three levels. The most local level is what CIPCA calls the Working Community (CDT). These working groups are created by CIPCA within the campesino community to be socioeconomic units that revolve around production. The goal is that all families in the community should join the CDT. By promoting collective forms of production the aim of the CDT is to incorporate production as an additional function of the local political and administrative units of organization (the community and the *sindicato*, a politico-administrative structure within the community). In educational activities, the CDT tries to foster openness to change by promoting a system of values based on participation, solidarity and equality.

The level of coordination above the CDT is the Union of Working Communities (UCDT). This guides the work plans of the CDTs, administers the pool of machinery to which the CDTs share access, and organizes the marketing of CDT products. It also supports technology transfer and offers technical training to CDT participants. The third level is the micro-region. This is the level at which development plans are conceived and implemented.

Production options for the working communities

In CIPCA's analysis, farm mechanization is essential for campesino development in the humid lowlands, where weeds and forest regrowth occur quickly. Without mechanization, the small producer is able to cultivate only a very small area – this generates insufficient income to sustain the family, with the result that members migrate further into the forest. Moreover, the improved technology that has been developed in the government research centre in Santa Cruz (CIAT) is adapted to the production systems of those medium and large producers who have access to machinery. For example, the higher-yielding varieties of rice that have been developed and disseminated require mechanized production systems.

Given this context, and on the basis of the results coming not only from CIAT but also from CIPCA's own experience and socioeconomic analysis, CIPCA decided to promote production options that revolved around the introduction of machinery to the CDTs. It considered this the only way for campesinos to be able to take full advantage of the technology developed in the research centres for medium and large producers.

The institutional organization of technology transfer in CIPCA

CIPCA has a Technical Department which designs experimental and demonstration plots, devises production packages, plans implementation, does follow-up work and organizes training courses.

In each area there are several field-work teams giving technical assistance for the production options that CIPCA is promoting. Each team consists of a social promoter, an agronomist, a vet and an assistant mechanic. They are responsible for the programme of transfer activities, for monitoring the experimental and demonstration plots, and for supporting the design of the communities' annual production plan.

CIPCA as a generator of technical information

From the outset CIPCA has attached great importance to agricultural research. However, it considers that conducting agricultural research should not be the responsibility of an NGO. However, in situations where there were no specialized centres available, CIPCA has undertaken research.

In 1976, when CIPCA began to work at Charagua in the region known as the Cordillera (an the area in the south of the department of Santa Cruz that borders the Chaco), there were no research organizations working in the area. CIPCA therefore decided to establish its own experimental centre in Charagua, where it conducted research into fruit-growing, pig and cattle rearing, and basic grains.

Some time later, however, CIPCA agreed to manage the centre jointly with CIAT, the government research institution in Santa Cruz. Subsequently the research centre became more fully integrated into CIAT's own institutional structure and is now one of CIAT's seven provincial Regional Research Centres (CRI) – centre's where the technology of the main experiment station is adapted to regional conditions. CIAT maintains a full-time technician at the CRI, and the centre's operating costs are shared between CIAT and CIPCA.

Similarly, CIAT and CIPCA have carried out on-farm trials collaboratively, sharing costs. To date, however, shared experiments have been limited to basic grains and stock-breeding, due to a lack of CIAT specialists in other areas.

In the colonization zone of San Julian, CIPCA carried out some on-farm trials with the CDT in areas where CIAT had not conducted research. The quality of the research, however, was poor and CIPCA felt that the results were not reliable. It therefore made a policy decision that it would only undertake experiments with campesino communities in cases where it already had a clear idea beforehand of what results to expect. The previous policy of conducting its own trials was abandoned in favour of a policy based on seeking better coordination with government research centres, and a policy of pressuring those centres to develop appropriate technology in the areas where CIPCA was working. Rather than go it alone, CIPCA therefore decided to make a concerted effort to draw down government to support its work. Direct inter-institutional contacts have facilitated this coordination.

CIPCA as an intermediate user of CIAT technology

Close knowledge of campesino circumstances and constant contact with technology-generation centres enhance CIPCA's ability to select technologies appropriate to the reality and capability of its target groups. In this role, CIPCA has become in many cases an intermediate user of CIAT technology. However, its function is not simply to gather information from research centres such as CIAT, but also to seek out institutions and request them to generate information specifically for CIPCA's projects. Coordination with such government institutions has not always been positive, either because of differences of opinion over what constitutes development, or because the target group of the technology generated was different from the CIPCA users. None the less, in other cases there has been successful coordination.

Linkage mechanisms between CIPCA and CIAT

CIPCA receives information from CIAT publications, courses and field days in the main experimental station. The participation of CIPCA's technical department in the annual planning meeting for CIAT's research also facilitates inter-institutional communication.

In the Cordillera, the close relationship between the CIAT technical officer at the CRI and the CIPCA technical team has been very positive. The technical officer informs CIPCA's team of research progress via monthly, yearly and three-yearly reports. In addition, the whole team makes two annual visits to the CIAT-run research centre.

Demonstration trials and testing technology

The CDT plays a central role in the testing and transfer of technology at a community level. Within each CDT there are individual paratechnicians who are given special responsibility for promoting local agricultural development and who therefore receive more extensive training. It is also with the CDT that CIPCA conducts demonstration trials. Some of the trials are run by CIPCA's technical officers, others are run by the community. The purpose of the trials is twofold: they demonstrate technologies, but in some cases they are also used to adapt experiment station technologies, bridging the gap between research results and the actual conditions of the campesinos.

The results from the demonstration trials run by the technical staff have been positive. In the Cordillera, for instance, trials were used to adapt maize technology, by increasing sowing density and reducing the number of seeds per planting. Among the settler farmers in Antofagasta, the trial aroused much spontaneous interest, and open days were organized. In several other communities, campesinos have begun to plant new winter crops, mainly potatoes and vegetables, as a result of CIPCA's demonstration trials. This spread of new technologies has also helped promote CIPCA's larger development model (of CDTs etc.) in this area.

Simultaneous to the trials that were run by CIPCA staff, the CDT in San Julian was also running demonstration trials, and indeed this was the main activity of the CDT in the early years of CIPCA's work in the area. In this type of trial the CIPCA technical team proposes a range of possible technologies, and the CDT chooses which alternative to test. CIPCA provides inputs, advice, a practical trial guide and training for the person responsible for agricultural development. The CDT carries out the experiment under the direction of this person. The technical officer and CDT members assess the trial periodically and prepare a report at harvest time to present to the community.

In some cases, after several years' work with demonstration trials, the campesinos, without help from the technical officer, have gone on to carry out their own experiments with new crops, new varieties and new crop-management practices.

Although both types of demonstration trial (technician led, and CDT led) have essentially achieved their objective, the trials run by the community itself have tremendous advantages:

- The campesinos are directly involved in looking for solutions to technical problems.
- New technology is combined with the empirical knowledge of the campesinos.
- Results can be directly replicated by the CDT.

The main disadvantage of this type of trial is that it can only be carried out by well-organized CDTs. In communities which lacked coherent organization, plots have been badly managed and have given poor results.

Training CIPCA's campesino clients

Initially, courses on specific subjects, such as handling agrochemicals, were held in every community and were open to all. To start with, although there were high levels of attendance at the courses, the practices that were taught were not put into practice.

Later, from 1987 onwards, more complex courses had to be given in response to the increasing complexity of technical problems in colonist farming systems. Because of this complexity, courses were more carefully targeted to the paratechnicians, who then had responsibility to pass on information in the communities. Once again, the results were poor, even if they had more impact than the earlier open-enrolment courses.

In the Cordillera, CIPCA has experimented with a new training system. Participants elected from the community have to attend four two-week technical courses over a two-year period. Teaching is done in the parish centre, and practical sessions are held at the experimental centre. At the end of each course an evaluation is carried out.

The results of this system have been more positive. However, the success of the courses depends on the existence of teaching facilities and the availability of good teachers. It is very hard to find people who have these specialized skills combined with teaching abilities.

A further form of training is for the campesinos to visit the research centres.

Conclusions

Over the years, CIPCA has come to appreciate its limitations in technology generation and has accepted that it should restrict itself to overseeing the last stage of technology transfer. This implies the need for close inter-institutional liaison and a clear division of labour between CIPCA and CIAT. Though not perfect, the two organizations have generated a number of mechanisms that help them coordinate and generate more-appropriate technologies. In this, CIAT concentrates on generating research results for CIPCA, while CIPCA concentrates on developing methods for effective transfer of the technology to farmers. CIPCA has experimented with a variety of methods for this transfer work.

PROFILE 4 El CEIBO

Bolivia

Central Regional Agropecuaria - Industrial de Cooperativas 'El CEIBO' Ltda (The El Ceibo Regional Agricultural and Agro-industrial Cooperative Coordinating Committee)

by

G. Trujillo

The case of El CEIBO, a federation of cooperatives in the Alto Beni of Bolivia, illustrates how a membership support organization oriented mainly towards cooperative development and marketing of agricultural products subsequently developed its own programme of agricultural technology development and transfer.

In the Alto Beni, public-sector research and extension programmes designed to meet campesinos' technological needs had by 1982 reached near-paralysis due to administrative, political and financial problems. However, farmers had realized the need for access to technical assistance and so El CEIBO was prompted to try and fill the gap left by the state. As El CEIBO developed its research and extension capacities, several factors helped its initiative. Among these were the existence of relevant technology and experience at government experimental stations upon which El CEIBO could draw, the enthusiasm of El CEIBO's promoters, which gave the organization credibility in the eyes of international NGOs, and El CEIBO's decision to seek out technical support from a whole range of national and international institutions.

CEIBO's area of work

The Alto Beni includes the provinces of Yungas North and South and Larecaja in the Department of La Paz. In 1961 the area was colonized in a programme directed by the Bolivian Development Corporation for Promotion (Corporación Boliviana de Fomento). Most of the settlers were Indians from the highland departments of La Paz, Oruro and Potosí. By 1987, there were some 7,000 settler families in the area involved in agriculture, with an average of 2.36 hectares of cacao per family; this remains the main commercial crop.

The origins and development of El CEIBO

In 1964, the Alto Beni Cooperative Ltd was founded by the government in a somewhat top-down fashion. As a consequence of administrative and political defects it achieved little success, and ceased to exist in 1971. However, the experience had highlighted to settlers the benefits of this type of campesino organization, so despite this initial failure of cooperativism, a more endogenous process of cooperative formation developed, and between 1971 and 1976 the first independent cooperatives emerged. These had two principal objectives: (1) to defend campesinos against the speculation of agricultural merchants in the area, particularly buyers of cacao, and (2) to prove that campesinos could run their own organization despite the failure of the Alto Beni Cooperative. The cooperatives organized a committee to coordinate the marketing of local produce. Although this achieved some success, the cooperatives recognized the need for a more formal central body with a broader scope of action. Consequently, in 1977 twelve cooperatives established the El CEIBO Cooperative Coordinating Committee Ltd (Central de Cooperativas El CEIBO Ltda), which today has thirty-six member cooperatives.

The founding objectives of El CEIBO were:

- to defend and represent member cooperatives and support these with agricultural inputs, education and the purchase and supply of foodstuffs and other household goods;
- to trade agricultural products in domestic and foreign markets;
- to obtain credit for production, marketing, trading, industrialization and services;
- to provide transport services and agricultural machinery;
- to process the area's agricultural products.

From the outset, El CEIBO had been under pressure to establish a special department with responsibility for technical assistance. However, as long as government extension agents were in the area, El CEIBO felt that it would not receive government approval to begin agricultural extension activities. Furthermore, campesinos did have access to some sort of technical assistance, albeit deficient in quality and quantity. However, in 1982 the strength of government research and extension services deteriorated rapidly, and they became unable to meet producers' technological needs. At this point El CEIBO began its own extension work and formed its own Department of Agricultural and Educational Cooperation (COOPEAGRO).

ATD in the Alto Beni: development and transfer of government technology

Between 1961 and 1977 the INC (National Institute for Colonization) introduced a substantial quantity of genetic material of cacao, citrus and coffee into the Alto Beni. As part of its programme of support to local settlers, it was involved in the production and distribution of seedlings to the colonists and established experimental/demonstration plots. Recommended technologies were based on experiences and skills which were unknown in the area. This was particularly so for cacao, which was a new crop. With hindsight, this policy of mass transfer without the back-up of local research created serious problems for the campesinos. The majority of the seeds distributed by the INC were based on a clone which was supposed to have high resistance to Witches Broom disease (*Crinipellis perniciosa*). However, after several years the disease began to damage farmers' cacao plantations. Furthermore, the clone had the disadvantage of producing very small fruit and stones, and over the years the INC-recommended clones have been abandoned by campesinos.

For several reasons – due to the technical failures of the INC as much as to changes in government policy – research and extension in the area was handed over in 1978 to the Bolivian Institute of Agricultural Technology (Instituto Boliviano de Technología Agropecuaria, IBTA) and was based in the Experimental Station at Sapecho in the Alto Beni. At the height of its work (1978–81) IBTA had six researchers and six extension agents at Sapecho. Their research produced useful results on cacao and other crops which has served, directly or indirectly, to improve campesino production. In addition, IBTA's presence led to visits from many Bolivian and foreign experts, who contributed to the knowledge of IBTA professionals and campesinos alike.

From 1981 onwards, however, IBTA went into decline, essentially due to financial restrictions. It could no longer meet the technological needs of campesinos in the area. Among the many problems that emerged in IBTA's work, the following were particularly serious:

- The high rate of turnover of technical staff due to retirement or change of jobs. This undermined the continuity of research and extension work.
- The technical staff came from the highlands and had little experience of tropical crops.
- Research records were lost, with the effect that the results of many trials were inconclusive.
- Campesinos lacked knowledge of IBTA's recommendations due to the extension agents' poor communication skills.

ATD in the Alto Beni: COOPEAGRO and ATD at El CEIBO

COOPEAGRO began life in 1982 with five promoters whose main activity was to combat Witches Broom disease in cacao through seeking technical solutions and training campesinos. Activities were coordinated with IBTA in Sapecho, with the aim of learning pruning techniques from IBTA's few experienced professionals. COOPEAGRO's initial results with campesinos were very positive. This experience showed the potential of using a membership organization as a conduit for technical assistance, and served to attract technical and financial support from international donors. In 1983 COOPEAGRO received funds from the Swiss bilateral assistance organization COTESU, to cover 60 per cent of its total budget, plus the full-time services of an agronomist sent by the German government agency DED as a technical consultant.

Furthermore, as COOPEAGRO's credibility grew, it received additional funding from El CEIBO – this funding was calculated as a percentage of the value of the cacao the cooperative marketed. This increase in funding paid for an increase in staff to seven paratechnicians in 1983 and the consolidation of a work programme with the following aims:

- to develop appropriate technology for cacao in the Alto Beni;
- to ensure long-term productivity and soil fertility;
- to train the COOPEAGRO team and agricultural promoters;
- to promote collective action by cooperatives and farmers in agricultural production and to foster an enabling cooperative ideology.

Following the IBTA model of technology development and transfer, COOPEAGRO established an experimental plot of ten hectares to conduct trials on cacao, and on systems of crop diversification. It was also used to produce seedlings for distribution. In its research work, COOPEAGRO has carried out trials on the following:

- family vegetable garden crops;
- use of chemical fertilizers to increase cacao productivity;
- selection of better cacao trees and trial plots;
- cultural methods to control Witches Broom in cacao;
- development of a package for biological production of cacao with marketing support from El CEIBO.

Agricultural extension methods are also conventional and similar to those of IBTA: extension revolves around visits to individual farmers, creating local user groups, talks, short courses, communal plant nurseries, demonstration plots, and so on. However, the one critical difference is that COOPEAGRO emphasizes the training of local campesino extension agents and paratechnicians – this seems to be the crucial reason for the greater willingness of farmers to accept its technical recommendations, in contrast to the failure of IBTA's extension agents.

As in El CEIBO as a whole, the guiding philosophy of COOPEAGRO is to promote the involvement of farmers in all its activities. Thus, the planning and evaluation of technology development activities is undertaken with farmers, which seems only sensible given that they are most informed about their own daily needs and problems.

Development of COOPEAGRO

COOPEAGRO has been flexible and ready to use periodic internal and external evaluations to improve its work. Initially, COOPEAGRO was divided into two sections, one specializing in education and the other in technical assistance. In 1985 it concluded that this led to excessive separation of the two components, as well as to resource duplication and poor coordination. The two sections were therefore merged and the twenty-two staff then underwent further staff training.

During the early years, staff rotated between different geographical areas and disciplines, but this tended to jeopardize follow-up of field and trial activities. The system has been modified to enable greater permanence of those staff who occupy critical and skilled positions in the organization. Indeed, it should be noted that COOPEAGRO gives much importance to the technical training of its staff, and helps develop their skills both through short trips as well as more substantial scholarships for study in secondary and higher education. Presently there are eight paratechnicians studying at the Universities of La Paz, Cochabamba and Santa Cruz.

Initially COOPEAGRO's staff were young and inexperienced, but they were also enthusiastic and dedicated. Many government institutions had little faith that COOPEAGRO would achieve anything because of its campesino membership and anticipated that it would inevitably make administrative, social and technical errors. In an effort to combat these negative attitudes, COOPEAGRO actively sought links with other institutions to win both the trust and the cooperation of other organizations. In 1989 it decided to hold bi-monthly meetings with other institutions to agree on the technical criteria on which to ground local ATD, and to maximize the use of the area's human and economic resources. Currently the bi-monthly meetings are attended by ten organizations active in the Alto Beni, and bring together public, NGO and farmers' organizations.

COOPEAGRO's relations with other organizations:

Having few technical staff, from the outset COOPEAGRO's policy was to seek help from other organizations whenever it needed specific assistance. Agreements with these institutions are sometimes formal, sometimes informal. Perhaps the most important single instance of external assistance was the presence of a volunteer technical consultant from a German development agency, who helped COOPEAGRO to develop and organize its approach to ATD.

Despite IBTA's budgetary problems, COOPEAGRO kept up close links with several of its experimental stations, using them as sources of technical information, staff training and genetic material.

El CEIBO also established formal relationships with several Bolivian institutions, such as the Agroecology Unit at the Cochabamba University (AGRUCO) and the Institute of Ecology (IE) of the University of La Paz (for assistance with natural-resource management and conservation). Technical assistance was also forthcoming from the Italian Technical Assistance Department (ACRA) for the establishment of cacao nurseries.

Despite being a campesino organization, El CEIBO also managed to develop close links with international organizations, including the Comissao Executiva do Plano da Lavoura Cacaueira (CEPLAC Brazil); the Colombian Agriculture Institute (ICA); the Tropical Agronomy Research and Teaching Centre (CATIE Costa Rica); the Inter-American Institute for Cooperation in Agriculture (IICA); Granja Luker (a Colombian chocolate company); and the National Federation of Cacao Growers of Colombia (FEDECACAO). The groundwork for these contacts was probably laid during IBTA's successful period in Alto Beni, before the formation of COOPEAGRO.

Conclusions

COOPEAGRO's performance illustrates that it can be viable for membership organizations to establish their own technology generation and transfer systems. In the case of COOPEAGRO, much of this success has been due to the following favourable factors:

- El CEIBO emerged from the failure of state-directed cooperativism, albeit not before the latter had shown campesinos the potential advantages of being organized. The failure of the Alto Beni Cooperative was due to the paternalistic approach of the state, rather than to a lack of campesino interest in cooperativism.
- The state's attempts at technology development and transfer, although apparently not very successful, did reveal the need for access to technical assistance and fuelled campesino demands for it. Furthermore, as COOPEAGRO got involved in extension, it recognized that there were existing technology and skills in the public sector waiting to be harnessed.
- Given the gap created by the shrinking role of the state, El CEIBO saw the formation of COOPEAGRO as a necessary and cogent response to campesinos' expressed needs.
- COOPEAGRO's seriousness of purpose and enthusiasm attracted finance and technical cooperation from international donors despite technical drawbacks.
- Through continuing analysis and evaluation, COOPEAGRO has been able to develop in line with changes in its own capacities and in local needs.
- El CEIBO and COOPEAGRO have always been active in seeking moral, technical and financial support from a broad range of organizations, initiating contacts rather than waiting for them.

ANNEX 2

PROFILE 5 CESA

Bolivia

Centro de Servicios Agropecuarios
(Centre for Agricultural Services)

by

A. Kopp and T. Domingo

CESA works in the high plateau and tropical areas of Bolivia. Like many other NGOs, CESA's early philosophy aimed at strengthening grassroots organizations. As it began to implement specific projects, CESA realized the importance of technology, particularly when it began to work in an area of recent colonization where there were no institutions generating technology. With scarce resources it was obliged to come up with its own strategy for establishing contacts with institutions generating agroecological technologies in other regions, and to make the best possible use of these contacts.

The origins of CESA

CESA began life in 1976 as ASEC (the Ecumenical Association for Field Work Coordination and Cooperation), an association set up by five churches to raise funds for development projects. Later it took on other functions, including implementing its own development projects. In 1983, political and other differences between the field team and the churches prompted ASEC to become an independent institution, CESA. Initially it concentrated on education projects. Gradually its support strategy changed; today CESA mainly undertakes agricultural projects.

The social imperative: early work in strengthening campesino organizations

The groundwork laid by ASEC provided CESA with an excellent network of contacts with community and cooperative organizations. At the outset CESA believed that low agricultural profitability was attributable to the state neglect of the rural sector, and that this could only be changed through organized campesino action directed against the state. CESA therefore devoted itself to strengthening campesino organization, but it had little idea of how to support the communities with agricultural technology. There are two projects which are worth analysing because the experiences with them have informed CESA's current way of working.

In the highland area of Sorata, CESA tried to re-establish the traditional Andean system of exchange and production relations between communities at different altitudes. CESA's team of agronomists promoted intercommunity projects in which campesinos from the 'valleys' (valleys leading down to the eastern lowlands) grew potatoes in the communal lands of highland communities. However, the lack of technical support meant that results were poor and hardly justified the physical effort involved. The initiative was therefore terminated.

In a second project among highland colonists settling in the tropical zone of the Alto Beni, CESA formulated a programme for organized groups of settlers to receive technical and financial support to help consolidate their new farms. However, the area designated for CESA's Alto Beni project was so remote, and of such low agricultural potential that the CESA team (which moreover had no specific training in tropical agroecology) was unable to contribute much to the settlers. Just as in Sorata, the inability to achieve significant improvements in production caused the campesinos involved to abandon the project.

These two examples are indicative of the work of CESA in its early years. The institution's political position and its concerns to strengthen grassroots organization were broadly shared by the campesino organizations; what was missing were the appropriate agricultural techniques. As far as CESA could see, other NGOs and state bodies had similar failings.

Recognizing the technological imperative: an agroecological project in Yucumó

As the settlers in the Alto Beni project abandoned their lands and moved further into the forest, so did CESA – to Yucumó, a tropical area in the south of the Department of Beni. State sponsored colonization began in Yucumó in 1983. The National Institute for Colonization (INC) gave land to groups organized in blocks of forty families – 25 hectares to each family and 180 for communal use. Presently there are roughly 1,500 families in Yucumó, mostly from depressed areas in the highlands.

The agricultural system in Yucumó involves clearance by slash and burn, followed by rice, the only marketable product. Due to lack of alternative techniques, the colonist is forced to clear a new area of 1.5 hectares annually; once cleared and cropped this land cannot be cultivated again. The campesinos do not usually sow winter crops.

CESA began its activities in the area by carrying out a fairly conventional diagnostic survey, with little campesino participation. A socioeconomic study highlighted the problem of accelerated soil exhaustion, and showed that in this system the campesino settlers were as poor as they had previously been in the highland areas they came from. A land-use study indicated that only 10 per cent of the soil is suitable for annual crops and that the greater part should be maintained as protected areas. In other words, the campesinos are sowing rice on land that is not suited to annual cropping.

Both CESA and the campesinos realized that the production system needed to be changed. Under pressure from the campesinos, and despite having no clearly defined alternative methods, a two-part production project on communal lands was launched, combining agriculture and cattle-breeding. In Yucumó there were neither state nor private institutions which could help in the design of these projects with expert advice on agroecological methods, and CESA had no funds to hire a full-time specialist. Therefore CESA was obliged to establish contacts with organizations working on tropical agriculture in other regions. Experts were invited on short contracts to advise the field team and to give practical training. These experts came from state institutions (CIAT-British Tropical Agricultural Mission, IBTA Chapare, the La Paz University Institute of Ecology, other NGOs and the Sub-Secretariat for Alternative Development – an institution responsible for coca substitution programmes). Links with state bodies have also been essential for the provision of seeds and seedlings for use in agroecological production systems. In addition to the visits of experts, groups of campesino promoters from the project have made study visits to experimental plots and stations at various institutions in Santa Cruz and the Chapare. Steadily, an informal network of contacts and experience has developed, greatly widening the range of tropical agroecology skills drawn on in CESA's project.

Originally the agricultural project proposed planting 30 hectares of perennial crops (coffee, cacao, citrus) after the first rice harvest. Over time though, as CESA developed its relations with other organizations and acquired lessons from them, the original proposal was transformed into a genuinely agroecological project. Among the main elaborations of the initial proposal have been:

- *Winter sowing* of pulses such as soya and beans. These crops have short-term advantages, that are both *economic* (they give a second annual harvest and improved diet) and *ecological* (e.g. by fixing nitrogen and controlling weeds, and so enabling a new rice planting).
- *Use of trees* in agriculture. This gives mid- and long-term *economic* benefits (fruit, firewood, timber, forage, green fertilizers, etc.) and *ecological* benefits (shade, fertilization, protection from erosion, etc.)
- *Analysis* of soil characteristics prior to clearing on the basis of the farmer's own knowledge and soil classification schemes.

In the cattle-breeding component, the project originally proposed improving the local breed of cattle to develop a resistant breed with higher milk production, along with the improvement of pastures and infrastructure. Here too the original proposal was amended with the introduction of new agroecological elements learnt from other institutions. These include alternative improved pastures, tree planting in the grazing lands, and the development of nurseries for multiplying pasture grasses.

Setting up a local network

As the project developed into an agroecological programme it stimulated much discussion amongst colonists and NGOs in the area. This led in early 1990 to the creation of an informal coordinating network called 'The Yucumó Interinstitutional'. It encompassed seven NGOs, a school, the parish, and colonists' organizations (*sindicatos*, or local unions, a women's organization, the paratechnicians' association and the school-leavers' organization).

The aims of the network were several: to simplify and standardize the terminology of the different organizations; to foster more rational use of the resources of each institution; and to coordinate research and training programmes. Committees have been created for agriculture, livestock and health-care activities. Each has responsibility for training promoters in its given area.

In July 1990 the network held a symposium in Yucumó on agroecology and invited experts from outside organizations. This has led to more systematic inter-zonal contacts via a programme of exchange visits. It also decided to put CESA in charge of the design and implementation of a participatory research project on agroecological production systems in Yucumó.

Participatory research into agroecological systems

The majority of the studies or research into agroecological systems, including CESA's, have failed in many areas: a lack of campesino participation, the abstract scientific nature of the work and its consequent lack of practical applicability. From the outset the new research project for which CESA had been given responsibility sought to avoid this by starting from campesino involvement. During the nine-month period of study – which at the time of writing was still running – two seminar-workshops will be held in fourteen communities of the area. In these seminar-workshops the colonists make a critical analysis of the economic and technical aspects of their production system. Their knowledge of soil quality is elicited, and the terms they use are documented and compared with more technical terminology. The experience of some colonists experimenting with improved land-use practices (introducing green vegetables, different pastures, etc.) is also analysed. The hope is that as a result, appropriate agroecological systems for the conditions of each community can then be identified.

The results of the seminar-workshops will be compiled in a teaching manual on agroecological systems, for use in extension work amongst the campesinos.

The projects implemented by CESA in the area are an important back-up to the study. They provide visible alternatives to the conventional and unsustainable model which small producers have usually employed, and which CESA had previously promoted in its earlier project in the Alto Beni.

As the study has not yet been concluded, it is obviously impossible to make any definitive assessment of its impact. However, the colonists' expectations have already been raised, as has their desire to work with new agroecologically grounded technologies. CESA has noted the following changes amongst the colonists since the onset of this more agroecological approach to development:

- Greater caution as regards clearing, and more careful selection of potential slash and burn areas.

- Requests for seeds for winter crops.
- Requests for trees to start up tree nurseries.

In local meetings of settler organizations, agroecological issues are now frequently raised, and farmers have given support to the paratechnicians' association and its work with ecological systems. In addition, base organizations and the Association have been lobbying other development institutions to give them training courses and genetic material.

Conclusions

In the early days of project implementation, CESA came to realize the important role that new technology had to play as part of a campesino development strategy. In Yucumó there were no institutions providing agroecological technologies, neither did CESA have the resources to bring in outside experts.

CESA therefore adopted a strategy in which it developed links with a broad spectrum of institutions, both NGOs and parastatals, working on technology generation in other tropical regions. The institutions sent experts on missions to Yucumó, and CESA's promoters went on study visits to the Chapare and Santa Cruz. Contacts with state institutions were particularly important because they provided a source of genetic material – the materials supplied now constitute part of the new technological options with which CESA works. With these links, CESA has managed to incorporate new technologies in its projects during implementation. Given that these technologies came from outside the region, the project became a testing ground for them, and so these other support institutions have been interested to monitor their performance.

The establishment of a local network enabled a broader and more effective discussion, dissemination and adaptation of these new practices. It hosted a symposium enabling other NGOs and campesino organizations of the area to establish contacts similar to CESA's with outside organizations.

A closing reflection

International resources channelled via public bodies have achieved good results in generating new technologies, but the benefits of these frequently fail to reach the small producer. Therefore the state–campesino–NGO triangle should be a future framework for optimum use of human, technical and financial resources. In the Yucumó area there is now an acceptable degree of coordination between NGOs and campesino organizations; it remains for the relevant public bodies to complete the triangle.

ANNEX 2

PROFILE 6 FUNDAEC

Colombia

Fundación para la Applicación y Enseñanza de las Ciencias
(Foundation for the Application and Teaching of Science)

by

J.E. Blandon and M. Prager

FUNDAEC was founded in 1974 by a group of teachers at a university in Cali (Universidad del Valle) who decided to orient their work towards the needs of campesino groups. The aim, clearly stated from the outset, was to 'seek new development options to generate forms of rural development that were relevant to all of society' and not only to its privileged few.

The Foundation centred its activities around implementing integrated processes of development in the following areas:

- Strengthening the regional economy.
- Improving rural education.
- Enhancing community organization.

Development and transfer of technology: public-sector institutions

In the words of the 1990 Colombian science and technology mission:

> Technologies based on the use of crop and livestock varieties which were highly responsive to the application of synthetic inputs and/or easy to mechanise led to the development of high yield, capital intensive systems which favoured large scale producers and provoked undesirable ecological consequences. (Cited in FUNDAEC 1991).

In general, these technologies have not brought about any improvement for the great majority of Colombian campesinos. Over time the state sector has begun to acknowledge this problem and develop new policies and proposals.

FUNDAEC's proposals for development and transfer of technology

With the benefit of hindsight, various stages can be identified in the development of FUNDAEC's approach towards science and technology. As we see below, for the founders of FUNDAEC, categories such as science, technology and technology transfer cannot be separated.

At its inception, FUNDAEC was convinced that the technology needed by the rural areas of the world already existed or was being developed in national and international research centres. On the basis of this position, FUNDAEC concentrated on devising educational strategies to bring modern science to the campesinos. While criticizing elements of the dominant conception of development (such as its rejection of traditional technology) FUNDAEC also (perhaps somewhat paradoxically) tried to transfer those elements of the modern technological packet that it saw, at that time, as beneficial.

As a result, FUNDAEC worked intensively in the first two years with a small number of families in the region, employing state-of-the-art technology 'packages' developed by national and international research. However, the results were far from encouraging, and the group began to see that much of the available technology was simply not adaptable to the precarious conditions in which most farmers in the region lived.

This initial enthusiasm for the Green Revolution packet was therefore followed by a second period in which the early failures led to a questioning, especially of modern agricultural technology. This questioning was accompanied by the vague idea that the campesinos' own traditional methods were probably the best-suited to their specific environments. This position shared common ground with some other NGOs.

This questioning led to a third period in which FUNDAEC sought to examine more systematically and deconstruct the process of technological development. They began by acknowledging both that industrialization-based models of development had not brought real progress for the majority of the poor, and that traditional technologies were unable to meet rural needs. This reflection led FUNDAEC to arrive at what was for them a new version of appropriate technology. Appropriate technology did not have by definition to be simple and cheap; rather, FUNDAEC argued, its 'appropriateness' had to be assessed in terms of its contribution to the well-being of the majority. The most pertinent questions to ask of any technology was therefore whether or not its introduction would improve the quality of life of a village, solve its main problems and bring greater freedom.

On the basis of this approach, FUNDAEC established a series of criteria to judge the appropriateness (or otherwise) of any given technology. Among the most important social criteria were that the population must be able to exercise effective control over the technology, and that the technology should contribute to the development of a technological and scientific culture in the rural community. By this second criterion FUNDAEC meant that an appropriate technology was one that the rural community would value not simply for its material impact in production, but because campesinos were able to understand and value the principles underlying the functioning of technology – technology had to be understood as well as simply applied. Overall, technology should increase the autonomy and food security of rural people, and not merely enhance output.

Elaborating systems-based technical proposals in FUNDAEC

FUNDAEC has aimed to elaborate technological options for campesinos on the basis of a systems approach to agricultural development. In much of its work, it has tried to adapt or introduce new sub-systems into campesino production. For FUNDAEC, the term 'sub-system' refers to a combination of different practices, crops and animals in a part of the farm. The NGO initially designed agricultural sub-systems on the basis of its knowledge of the region and the needs of campesinos in the North of Cauca. These sub-systems constituted production alternatives.

When FUNDAEC transfers these alternatives, the intention is not simply to transfer the technologies *per se*, but more importantly to impart an understanding of the principles underlying the operation of the sub-system so that campesinos might be able to use those principles in their own experiments and farm management decisions.

In addition to research on alternative production systems, FUNDAEC has carried out complementary research on the following topics:

- Livestock nutrition. The main focus has been on seeking various alternatives to the use of concentrated feeds, in particular they have looked at the possibility of feeding livestock with plants found in the sub-system, and the feasibility of local concentrate production based on cassava.
- Soil management and conservation. The conservation techniques with which FUNDAEC has worked include the maintenance of continuous crop cover, the incorporation of micro-organisms to improve the efficiency of fertilization, and the use of organic fertilizers and crop associations as responses to acidic soils (the majority of soils in the region are acidic).
- Pest control. This work has involved the introduction of trap-crops, biological control techniques and crop associations to reduce pest damage.

Techniques for seed production within the sub-system are also being researched.

Other research projects carried out by FUNDAEC are related to the development of rural agro-industry. Techniques have been developed for processing fruit, soya milk, cacao and concentrated livestock feeds.

Aside from introducing a new set of practices, this work with sub-systems is meant to be a basis for strengthening community organization. In research and transfer actions, FUNDAEC aims to foster campesino collaboration in a series of local development activities.

Formal education: the best strategy for technology transfer

The development of a formal, out-of-school based teaching system (the SAT) began in 1980. The programme has since been recognized by the Ministry of Education as an approved experimental programme for the North of Cauca, and has since been incorporated into the formal agricultural courses in technical colleges.

The SAT offers three training levels: promoter, technician and bachelor's in rural welfare. The first level – promoter of rural welfare – is the basis of the system. It aims to consolidate in young people the abilities, knowledge and attitudes required for service to the community. The course includes components of agricultural technology, mathematics, science and languages.

The Ministry of Education has recognized the course up to the level of promoter as equivalent to the first two years of secondary education. Its duration depends on the time that each group and student devote to it, but it generally takes students between three and four years to complete it. The two subsequent levels have been developed with similar criteria and in the same basic areas of skills development.

The curriculum is developed at a field level, and teaching is organized in local communities. Tutorial meetings are guided by instructors trained by FUNDAEC. Much emphasis is placed on practical work and research, and the majority of activities are carried out in the countryside and require no special installations or equipment.

In several parts of the course – not only those related to agricultural technology – students analyse and apply the knowledge that has been generated by FUNDAEC's own research into campesino production. For example, unit two of the agricultural technology module at promoter level contains part of the theory developed by FUNDAEC on agricultural sub-systems. It also encourages students to establish and experiment with agricultural sub-systems of their own.

This unit does not just give students definitions of concepts, but rather encourages the students to construct concepts themselves. These can then be tested in new situations. The aim is to facilitate the transfer of the logic of the sub-system, as well as the transfer of the technology itself.

Subsequent levels of the programme focus on issues such as the development of proposals for alternative sub-systems, crop diversification and farm management.

General relations with the public sector and other institutions

FUNDAEC explicitly aims to collaborate with a small number of public or private institutions working on problems of small-scale agricultural production, in order to share experiences and – if resources permit – begin to carry out joint research and application projects in regions other than the North of Cauca. These interactions are both informal and formal, local and national.

Some of FUNDAEC's inter-institutional relations are occasional interactions which occur at specific times and serve to exchange information and disseminate methodological techniques. FUNDAEC often presents the results of its research at conferences in universities and in both public and private institutions. Such interaction has sometimes generated more significant collaborations, such as networks of NGOs which share research results and field experiences.

As regards relations with international bodies, FUNDAEC has maintained links with the International Centre for Development Research for a decade, especially in the area of research on alternative production systems.

At national level, FUNDAEC participates in a rural educational project with the Ministry of Education that is part of a Latin American programme aimed at improving basic education. Colombia chose as its focus the improvement of rural post-primary education and the development of practical methodologies for basic education in marginal urban areas. FUNDAEC is part of the coordinating committee for rural post-primary education, along with one other NGO and various local education institutions.

Through an agreement with the Ministry of Education, the SAT programme is being implemented in nine Colombian departments through agreements with different governmental and non-governmental institutions. FUNDAEC also sustains quite close relations with technical colleges. The link with universities is particularly strong, aiding research collaborations and exchange of information. These links allow university academics to have closer contacts with field realities, while the NGO gains access to laboratory facilities, experience and research teams.

A specific case of FUNDAEC–state interaction: the MONDOMO collaborative project, Cauca

The MONDOMO collaborative project brings together several institutions which are all working on agricultural development in the department of Cauca. These include a producers' organization (the Cattle-Ranchers' Fund of El Valle), an international agricultural research centre (the International Tropical Agriculture Centre, CIAT), a state development agency (the Autonomous Regional Corporation of the Valley of Cauca, CVC), the NARS (the Colombian Agricultural Institute, ICA) and FUNDAEC.

The project was started in 1986. It covers 100 farms, but this number is set to increase. The area permanently attended covers some 1,116 hectares. The broad aims of the project are:

- to conduct research into, and disseminate agricultural sub-systems appropriate to agricultural, social and cultural conditions in the zone;
- to make available to farmers pasture technologies developed by CIAT;
- to evaluate and document methodologies employed in the diffusion of forage grasses and vegetables.

The institutions decide the form of their respective participation in the project in a meeting which is held at the start of each phase of activities. Participation depends on their respective abilities, experience and interests. In this sense, the project aims to exploit potential complementarities between the different expertise of these institutions. In this particular project, CIAT's strength is advanced agricultural research into production systems, particularly on acidic soils; ICA draws on its long history of crop research in the Cauca; the CVC on its experience of soil conservation work in the upper valley of the river Cauca; and FUNDAEC on its expertise in training groups of producers, in participatory research methods, in systems research and in the management of credit programmes for campesinos.

The project has several technical foci. One aspect of the project has been to promote systematic reforestation. The project promotes reforestation alternatives which will provide forage crops for animals, aid soil recovery and also provide an economically attractive option for farmers. Also, at present about thirty perennial forage crop species are being promoted amongst campesinos in the region; the results look promising. Another line of action of the project is soil recovery based on sowing leguminous pastures and crop varieties suited to the region. The proposed aim is to achieve a rapid and efficient soil coverage with species which contribute to soil conservation or soil fertility.

These activities were begun on ten farms in the region whose experiences have been closely followed. The results have been organized in a database which is constantly updated. At a later stage, plants were distributed so that campesinos could repeat the experience by themselves, with monitoring by the project.

Promising results are being obtained with sub-systems based on cassava, a crop which campesinos grow because it resists drought and because of the demand for cassava from starch producers.

Overall, this collaboration has on the one hand allowed FUNDAEC to make a contribution to a large collaborative activity, and at the same time has facilitated its access to a wide body of institutional expertise.

Lessons learnt and recommendations

FUNDAEC's experience leads them to conclude that links between NGOs and public-sector institutions are desirable in order to obtain greater impact and efficiency in programmes aimed at the campesino sector. However, some difficulties remain.

Circumstances which, from FUNDAEC's point of view impede this collaborative process are summarized below:

- Frequent changes in government policies from one administration to the next, frustrating the institutionalization and continuity of social development programmes.
- The Colombian government's policy of market liberalization.
- The continuing fall in resources allocated by the government to the poorest rural sectors.
- The relatively unstable position of decision-making officials, which means that even under the same government, there are many changes in the development of a project.

In order for NGOs to meet this challenge successfully, they must coordinate their actions through mechanisms such as information exchange networks, and collaboration on projects. Relationships with some public-sector institutions, such as universities, are more mutually rewarding than others.

Differences of opinion between FUNDAEC and the public sector will never cease to exist, but none the less there are real potential complementarities between the two that can be exploited to the benefit of the rural poor.

PROFILE 7 IDEAS

Peru

CENTRO IDEAS
(The IDEAS Centre)
by
J. Chávez Achong and A. Srecher Snauer

Institutional origins and perspectives on development

IDEAS was established in Peru in 1978 as one of a number of private development organizations set up during a period of popular organization against the military dictatorship when the state-planned structural reform process had ground to a halt in 1974/5. IDEAS was founded by people linked to left-wing parties or others involved in pastoral work in both Catholic and Protestant churches, and who had worked on some non-governmental development projects in the 1960s and 1970s.

The evolution of IDEAS as an institution has passed through at least three stages: the first was a period in which activities concentrated on popular education and support for land-poor campesinos and marginal urban groups. In its second stage, IDEAS moved towards becoming a development institution which laid emphasis on productive activities and on coordination with other NGOs. During the third stage of its evolution, it broadened its role – in two senses: it began to work with other social groups, and it started to develop policies with the goals that they be adopted by both civil society and the state.

In the first period the NGO saw itself in opposition to the state. It aimed to support popular organizations through a broad combination of research activities, education and services. In the second period, it adopted more specific approaches such as ecological agriculture and micro-regional development, and efforts were made to professionalize the team. In the third stage, IDEAS has been more concerned with technological factors and development management at a local level. It has begun to work with concepts from agroecology and micro-drainage basin planning, at the same time as it has been more explicitly concerned to promote collaborative relationships between different social, private- and public-sector organizations. The aim is to develop a shared vision of development and combine different disciplines and expertise.

In some respects, then, IDEAS currently operates in several spheres:

- as a socioeconomic and technological research institution;
- as a training and extension institution;
- as a political actor, seeking (as it believes NGOs ought) to influence decision-making and to contribute to the democratization and modernization of the state.

The challenge is to link the work it conducts in these different spheres. Theoretically, the research is meant to contribute to its training work, and research and training together are supposed to generate policy ideas and methods that could be offered as suggestions for public-sector policies and programmes. IDEAS however recognizes that these linkages are not as strong as they ought to be and in the future is concerned to improve the links between research, extension training, and policy debate.

Experience of technology development and transfer

IDEAS works in three parts of rural Peru. Two of those projects are analysed here: the first was related to ecological agriculture in the northern Peruvian highlands, the second involved campesino agro-industry in the central highlands.

Ecological agriculture: the San Marcos-Cajamarca Programme Through a series of projects, the IDEAS Cajamarca programme in the north-eastern region of Marañon has supported families and campesino organizations by raising production levels, diversifying and preserving productive resources, and improving the quality of life, particularly in the areas of health and nutrition.

IDEAS owns a demonstration plot of less than 2 hectares in marginal lands of the campesino village Penimpampa. This has allowed it to put agroecological principles into practice and adapt them to conditions in the zone. This demonstration plot served as the point of departure both for agricultural experimentation and for extension and training work aimed at campesino families in various villages. Subsequently, the work has been extended to dozens of families in the micro-region. In addition, the project has one plot and two horticultural allotments in the city of San Marcos. These allow experiments in urban agriculture that can be used as the basis of research to develop proposals to put to the urban population and to the municipal authorities.

The technological basis of the ecological agriculture programme includes the following:

- Regeneration and conservation of natural resources, including soil, water, germplasm, and flora and fauna.
- Management of productive resources, which includes crop diversification,

recycling of organic materials and nutrients, biotic regulation and cultivation and animal husbandry techniques.

- development and management of micro-climates and improvements to housing.

The target groups for the programme's training and extension work are campesino organizations. Collaborative ventures with public- and private-sector institutions have largely been informal; they have been based on establishing common objectives and finding people with the will to cooperate.

The results of the Cajamarca Programme over the past eight years can be summarized as follows:

- IDEAS has developed a detailed knowledge of the zone and positive cooperative relations with villages and campesino organizations.
- It has managed to develop a demonstration plot for ecological agriculture located in conditions less favourable than average campesino conditions.
- There have been improvements in agricultural production on a growing number of campesino plots thanks to the diversification and extension of vegetable cultivation, and the adaptation and application of new techniques.
- The spread of environmental improvements. These are reflected in housing; reservoirs; systems for distributing drinking water; health and nutritional standards; and the increased use of resource-conservation technologies.
- campesino organizations have been strengthened and have been placing increasing pressure on the state to provide basic rights. This observation is linked to an achievement which, although very difficult to prove, seems important. By strengthening these rural organizations, IDEAS has helped strengthen the capacity of civil society to resist forces of repression that stem from the simmering civil war in the Peruvian countryside.

Among the problems which have been identified are the differentiated effects of extension and training between men, women and children; the failure to integrate the health and nutrition work into the productive dimensions of the project; and the continuing uncertainties as to whether ecological agriculture is economically viable.

Agro-industry: the IDEAGRO plant In mid-1985, IDEAS decided to enter the field of agro-industry because of its potential as a bridge between rural and urban development programmes, and its strategic importance for national development and food security. The agro-industrial programme IDEAGRO was set up around a project to develop industrial plants to process rural products into food-mixes. The programme was based on a proposal that CIP, the International Potato Centre, had developed. The plant is located in a zone of the Mantaro Valley. This was chosen because of its access to raw materials and to markets, the local availability of labour and services, and the history of agro-industry in the region.

The aims of the programme are to investigate the problems of small-scale food-processing for domestic consumption; to develop agro-industry as a support to local campesino production and marketing efforts; and to contribute to the elaboration of national policies for agro-industrial development.

The production plant concentrates on the processing of cereals, vegetables and root-crops. The processes include selection, boiling, drying, toasting, fine and coarse milling, along with the mixing of flours.

IDEAGRO builds on the techniques and processes characteristic of cottage industry and of small-scale Peruvian agro-industry. It develops these according to modern marketing criteria and by incorporating new techniques. The products have to meet both the different needs of different consumers (regional, traditional, modern urban and institutional) as well as satisfy certain technical and quality criteria.

In order to develop appropriate agro-industrial technology, great effort was made to ensure that the plant used machinery produced within Peru, and some of it in the region itself. Only the milling stone and a packing machine are imported. IDEAGRO has developed a new type of toasting machine and a drying machine based on solar power which provides a complement to the conventional drying machine.

IDEAS set up IDEAGRO as its own (rather than a campesino) firm because of the initial economic risks involved. It concentrated its early efforts on trying to make the plant competitive in a free-market context. Ultimately the idea was that if the technology was competitive it might be transferred elsewhere through the market. The problems of how the ownership of such plants should be structured will be dealt with once the issue of economic viability has been resolved.

The results of the IDEAGRO project can be summarized as follows:

- Machine design has been improved as a result of the IDEAGRO experience.
- Both in Lima and in the central highlands, product lines produced by the plant have begun to appear as bottled and canned products in supermarkets.
- The firm is now seen both by governmental institutions and national and international NGOs as a valid intermediary.

However, various factors, both internal and external have led to a temporary paralysis of these activities, and have not permitted the replication of the experience. These obstacles – which have undermined the viability of the plant – stem primarily from the economic chaos in Peru and from problems of rural violence.

Relationships with the public sector and other development agencies

Relations with the public sector After a period of distant and strained relations between NGOs and the state which began in the 1960s, NGOs in Peru have begun to take a more prominent role in national affairs as result of the deepening crisis of the Peruvian state and the quantitative and qualitative growth of NGOs.

In IDEAS' experience, state–NGO cooperation, although desirable, is difficult to implement, unstable and unevenly developed in different regions of Peru and different sectors of economy and society. At a general level, NGO–public-sector relations can occur through the establishment of coordination mechanisms for programme implementation, and in the definition of macro-policies. In the former, both sides participate in implementing activities, but in the latter the NGOs have little genuine participation. The further problem for left-leaning NGOs in Peru is that they find themselves caught between a state whose military apparatus distrusts them and suspects them of being subversive, and a terrorist movement, Sendero Luminoso, which considers them to be reformist and indeed has killed NGO staff.

Notwithstanding the difficult context, IDEAS has established formal coordinating links with national universities, with regional development projects, with the experimental centres of the NARS, with local and regional governments and with funding institutions.

Collaborative relationships with NGOs IDEAS collaborates with five different NGOs, with the aim of merging some areas of work and discussing the experiences of different NGOs on subjects such as technology, training, environment, and promotion of small-scale industry. At an international level, IDEAS is a member of the Consortium of Agroecology and Development (CLADES), of the Rural Agro-industrial Development Programme (PRODAR) and the World Federation of Organic Agriculture (IFOAM) which includes NGOs and organic producers from the North and South. Finally, in each of the regions of Peru in which it operates, IDEAS participates in local and regional coordinating bodies which bring together NGOs and decentralized public-sector programmes. It actively participates in establishing local management bodies which facilitate coordination between grassroots organizations, local government, church groups and NGOs.

PROFILE 8 CAAP

Ecuador

El Centro Andino de Acción Popular
(Andean Centre for Popular Action)
by
CAAP

Institutional origins and perspectives on ATD

The Andean Centre for Popular Action (CAAP) was established in 1978 in the context of currents of critical Ecuadorian thought on Indian and rural politics.

These discussions led to a decision on the part of CAAP's founders to investigate the economic, sociopolitical and cultural conditions of indigenous Andean *campesinos*. Until this point, the production techniques and rationalities of indigenous campesinos were little understood in Ecuador: CAAP aimed to fill this gap.

According to CAAP, this was important because of the weaknesses of the Green Revolution programmes of Ecuadorian public-sector institutions, such as INIAP. These programmes demanded a form of social and economic organization and an agroecological context which did not match the realities faced by Andean campesinos. In addition, argued CAAP, they did not allow real campesino participation, nor did they take account of the structural conditions behind the growing crisis in Andean agriculture (such as lack of land; demographic increase; rural–urban migration, etc).

CAAP's approach to ATD builds on the argument that campesino production is based on certain agronomic principles which 'modern' research should enhance and strengthen: a position they contrast to that of those (such as INIAP) who argue that campesinos need new production approaches based on modern technology. However, CAAP argues that it tries to build on current agricultural practice and not on some unattainable 'Andean ideal' – although there are some in CAAP who criticize some staff for being over-critical of INIAP and for being too romantic about the value of traditional technology.

Thus, CAAP proposed an alternative form of development for campesino agriculture based on the logic, objectives and realities of campesino production. The institution spent many years working with campesinos in the northern highlands of Ecuador to understand the logic and objectives of their systems of production. On the basis of this work, CAAP identified three main objectives of Andean campesinos: the long-term preservation of the ecosystem; the provision of food to satisfy family nutritional needs and the demands of social, cultural and ritual practices; and the optimal use of available labour. It stressed that any technology research programme should respect these objectives as part of the structural context of campesino agriculture.

ATD at CAAP: an NGO's experience with an experimental farm

As CAAP began its programme of ATD, it tried to base it on three main principles:

- A form of campesino participation that implied working on the basis of indigenous knowledge, and campesinos' own production concerns.
- A methodology that was as far as possible, scientifically rigorous.
- An inter-disciplinary approach to research – on the grounds that it was essential to consider the cultural, social and technological dimensions in Andean agriculture together.

The first step in the research entailed promoter-observers in the countryside maintaining continuous contact with the campesinos of Cayambe. This ethnographic approach, using participant observation to collect information, was combined with more formal methods of social research, such as questionnaires, group meetings and use of secondary material. These different data sources were used to construct an understanding of the conditions and logic of Andean production systems. They also generated knowledge of the main technical problems faced by campesinos.

Initially, CAAP sought solutions to these problems through on-farm research. However, this met two problems: campesinos saw the packages as risky; and researchers were unable to ensure that they gathered reliable data. For these reasons, CAAP decided to buy land and set up an experimental farm in Cayambe to carry out experiments in controlled conditions before beginning on-farm trials with campesinos. To ensure the relevance of these experiments for the campesinos in the region, CAAP did not buy flat, irrigated and fertile land (i.e. the type of condition characterizing INIAP's experimental stations). Instead, it bought a piece of land in an area of campesino families which reproduced the agroecological conditions found on campesino plots. The farm was located on sloping, eroded and rain-fed land.

From the outset, work on the farm was based on the following research themes:

- Research into local technologies to explore the possibilities of enhancing their effectiveness.
- Experimentation with Andean crops, above all crops from other Andean regions, to assess how adaptable they were to the northern Ecuadorian Andes.
- Experimentation with soil conservation on slopes.
- Experimentation with levels of fertilization, with the aim of promoting the use of organic fertilizers.
- Pest control through crop rotation and use of traditional methods and practices;
- Domestic animal husbandry.

This initial programme had several limitations. Although some information on Andean technologies was collected, analysis of these techniques was poorly systematized and rarely quantified or subjected to cost–benefit analysis. It was also methodologically flawed, in that the impact of certain interactions (such as that between changes in soil fertility and the incidence of pests and plagues) were overlooked. Perhaps the greatest error at this stage, however, was that despite the supposed concern to do research, much of the initial work on the farm became an effort to demonstrate terracing and other soil conservation technologies, which were moreover technologies from outside the region. In this regard, CAAP contradicted its own argument that a locally grounded Andean approach to campesino agriculture had to be sought and that careful research had to precede demonstration. This mistake became evident in termly meetings with campesinos, who stated that they could not terrace their land and that therefore CAAP's work was of no use to them.

The apparent reasons for this failing are instructive. Although CAAP's rhetoric stressed the imperative of careful research, the institution still attracted many who were concerned to 'do something' for campesinos. It also attracted many who believed in certain agroecological technologies. Ultimately these people became frustrated with research and began wanting to demonstrate and transfer technology.

Recognizing these mistakes and the other methodological limitations of the research, the focus of CAAP's work on the farm was changed. On the basis of a regional survey of production systems, rotations and existing varieties in the parishes of Juan Montalvo and Cangahua, a decision was taken in 1986 to concentrate work on the farm around the three challenges that seemed paramount to campesinos. These were:

- how to increase production;
- how to maintain soil fertility;
- how to control pests.

On the farm, efforts have been concentrated on an analysis of the impact of crop rotation, varietal selection and cultivation techniques on these three goals. Research has been based on the technologies identified in the initial survey of farmer practices. Thus the soil fertility implications of the cereal–potato–legume rotation have received much attention, as have technologies for the cultivation of barley, potatoes, wheat, beans, maize, peas and onions. Various changes were introduced and compared with the campesino production system. New crops were introduced to measure the effects on soil fertility (in the case of legumes) and on income (in the case of rye). In addition, CAAP began work with indigenous tree species.

In the research on rotations, through careful management of the limited space available, it has been possible to devise a design which generates information both on short cycles (potatoes–cereal; cereal–legumes, etc.) and on longer cycles. However, this work has faced problems: the analysis of the productive and economic impact of the different rotations takes several years and there is considerable risk that climatic factors will influence the results. Consequently, even after four agricultural cycles CAAP still lacks definitive data. However, at the very least it has become clear that the general practice of crop rotation has had a positive impact on productivity and soil fertility on the farm. In this sense the farm has once again served a demonstrative function for neighbouring campesinos.

In the research on crop varieties, the objectives were to:

- maintain the genetic diversity of traditional crops by bringing in varieties from different regions;

- evaluate these varieties under conditions of low-input campesino production; and
- give campesinos access to a greater number of varieties.

This work differs from that of INIAP because INIAP does not evaluate different species under conditions of campesino agriculture, and does very little research on native varieties.

First, different varieties were collected, then preliminary tests were carried out to obtain basic data about their different characteristics and behaviour. Subsequently, comparative statistical tests were carried out, and lastly seeds of proven varieties were multiplied on campesino plots. In all, 135 samples were collected, covering sixty-three distinct identifiable varieties of thirteen crops. The research concentrated on the two main crops of the zone: potato (twenty-eight samples, seventeen varieties) and barley (thirty-seven samples, eleven varieties).

The limitations of ATD at CAAP

Although CAAP's research has filled gaps in knowledge about technologies and methodology in Ecuador, it has come up against important obstacles. There are still methodological problems in the management of multiple data over time and space, and even now CAAP's researchers do not fully understand the implications of much of the data they have accumulated. CAAP recognizes the lack of rigour in its research, a result of its institutional weakness in certain areas, such as statistical analysis. It has acknowledged its need for methodological support from other public institutions to solve these problems. Although it has not sought such support formally, it has hired INIAP staff as consultants to assess CAAP's research.

Inter-institutional communication

General communication with other institutions about research and availability of technologies has been inadequate. Information has been exchanged with other NGOs, but on the whole the results from the experimental farm have stayed within CAAP. Moreover, other institutions such as CESA see any results CAAP may have as too theoretical and still unproven. In order to tackle this lack of communication and to stimulate internal discussion, the creation of a specialized office within CAAP responsible for communication has been suggested. However, resources are not available for this, and so no such office has been created. None the less, within CAAP some staff favour greater contact with public institutions and NGOs to disseminate more widely the results of the research done on the experimental farm.

ANNEX 2

Technology transfer

Until now, little of the technology developed on the farm has been taken out to the countryside. All transfers that have taken place have been of proven varieties, not of crop rotations (a project which may well take ten years before producing validated recommendations). This limited transfer work reflects CAAP's insistence that technological proposals be coherent and proven before they are disseminated; until now, the results from the crop rotation trials have been very preliminary.

CAAP's resistance to any 'transfer' actions also reflect its conception of technology transfer, which rejects the validity of the orthodox extension practices of 'transferring new technology' and of introducing new institutional practices into the community. Rather, CAAP wishes to enhance and strengthen existing campesino knowledge and to respect traditional forms of communication so that these forms become the means of any communication of CAAP's results. In addition, CAAP does not believe that it is possible to transfer technological packages in such a varied environment as that of campesino agriculture. Therefore, as regards 'extension' CAAP's work is restricted to (1) discussing technological ideas with campesinos in termly meetings on the experimental farm and through informal contacts, and (2) the selection and production of seeds which allow campesinos to carry out their own tests, make comparisons and comments, and to feed these back into research on the experimental farm. The principle underlying this approach is that campesinos will adapt the ideas and inputs they get from CAAP, although even in this regard impact is constrained because the 7 hectares of the experimental farm does not produce sufficient seed to meet local demand.

It should be emphasized that there is an ongoing debate within CAAP about these conceptions of technology transfer. The overall lack of technology transfer has resulted in tensions between those who want to carry on with the research and those who feel that the limited impact in the campesino economy which the farm has had to date does not justify the time, personnel and money that the experimental farm requires. The early tension between 'research' and 'action' persists.

ANNEX 3

VARIOUS DEVELOPMENT INDICATORS

Bolivia, Chile, Colombia, Ecuador and Peru

Macroeconomic indicators

Gross national income per capita (1987 US $)[a]

	1970	*1975*	*1980*	*1985*	*1987*	*1988*	*1989*	*1990*
Bolivia	740	820	830	610	550	530	540	600
Chile	1,580	1,090	1,510	1,200	1,390	1,520	1,630	1,870
Colombia	760	890	1,070	1,040	1,100	1,080	1,130	1,130
Ecuador	720	1,020	1,320	1,170	990	1,060	1,020	1,010
Peru	1,300	1,400	1,410	1,170	1,360	1,190	1,080	1,050

Gross domestic product (average annual growth %)[a]

	1970	*1975*	*1980*	*1985*	*1988*	*1989*	*1990*
Bolivia	0	7	−1	0	3	3	4
Chile	2	−13	8	2	7	10	4
Colombia	7	2	4	3	4	3	4
Ecuador	6	5	5	4	11	0	3
Peru	5	3	5	2	−9	−10	2

Total External Debt as a % of GDP[b]

	1980	*1985*	*1988*	*1989*	*1990*
Bolivia	93.3	176.6	117.9	96.9	101.0
Chile	45.2	143.3	97.1	76.0	73.6
Colombia	20.9	42.9	45.2	44.7	44.3
Ecuador	53.8	77.4	115.2	120.5	120.6
Peru	51.0	90.8	104.3	73.5	60.1

ANNEX 3

Aid receipts as a % of GDP[c, d]

	1980–1	*1984–5*	*1987–8*	*1989–90*
Bolivia	6.1	7	8.9	12.2
Chile	–	–	–	0.3
Colombia	–	–	–	0.2
Ecuador	–	–	–	1.5
Peru	1.1	1.7	0.8	1.5

Agricultural sector

Percentage of labour force in agriculture[d]

	1970	*1975*	*1980*	*1985–8*
Bolivia	52	49	46	46.5
Chile	23	20	16	18.3
Colombia	39	37	34	–
Ecuador	51	45	39	38.5
Peru	47	43	40	35.1

Exports of non-fuel primary products as % total exports[e]

	1970	*1975*	*1980*	*1985*	*1987*	*1988*
Bolivia	92	67	73	40	29	42
Chile	96	89	89	90	90	84
Colombia	82	72	77	65	46	49
Ecuador	98	37	34	30	55	49
Peru	98	94	63	57	70	71

Percentage of GDP accounted for in agricultural sector

	1960	*1981*	*1985*	*1989*
Bolivia	26	18	27	32
Chile	9	7	–	–
Colombia	34	27	20	17
Ecuador	26	12	14	15
Peru	18	9	11	8

Growth in agricultural production (%)[f]

	1960–70	*1970–81*	*1980–89*
Bolivia	3	2.9	1.9
Chile	3.1	3.0	4.1
Colombia	3.5	4.7	2.6
Ecuador	–	2.9	4.3
Peru	3.7	0.3	3.6

ANNEX 3

Social indicators

Life expectancy at birth (years)[a]

	1970	*1975*	*1980*	*1985*	*1987*	*1988*	*1989*
Bolivia	46	48	50	52	53	53	54
Chile	62	66	69	71	72	72	72
Colombia	61	63	66	68	68	68	69
Ecuador	58	60	63	65	65	66	66
Peru	54	56	58	60	61	62	62

Notes:

[a] World Bank; World Tables, 1990 update (on diskette).
[b] OECD; Development Cooperation Report, various editions.
[c] OECD; Development Cooperation Report, various editions.
[d] UNDP; Human Development Report 1991.
[e] Calculated from World Bank; World Tables, 1990 update (on diskette).
[f] World Bank; World Development Report, various editions.

REFERENCES

The asterisk marks papers that were prepared for the research and then presented at the workshop: 'Taller Regional para América del Sur. Generación y Transferencia de Tecnología Agropecuaria; el Papel de las ONGs y el Sector Público' ('Regional Workshop for South America. The Generation and Transfer of Agricultural Technology: the Role of NGOs and the Public Sector'), held in Santa Cruz, Bolivia between 2 and 7 December 1991.

Adriance, J. (1992) 'User Friendly Agroecology Networks', *Grassroots Development* 16(1): 44–5.

AGRARIA (1991) *Un Desafio al Desarrollo: Experiencia de Apoyo al Pequeño Agricultor*, Santiago: AGRARIA.

Agudelo, L.A. and Kaimowitz, D. (1991) 'Institutional Linkages for Different Types of Agricultural Technologies: Rice in the Eastern Plains of Colombia,' *World Development* 19(6): 697–703.

* Aguirre, F. and Namdar-Irani, M. (1991) 'Relaciones Sector Público-ONG. El Caso de AGRARIA.'

—— and —— (1992) 'Complementarities and Tensions in NGO–State Relations in Agricultural Development: The Trajectory of AGRARIA, Chile', *Agricultural Administration (Research and Extension) Network Paper 32*, Overseas Development Institute, London.

Altieri, M. (1987) *Agroecology: The Scientific Basis of Alternative Agriculture*, Boulder: Westview Press.

—— (1990) 'Agroecology and Rural Development in Latin America', in M.A. Altieri and S.B. Hecht (eds) *Agroecology and Small Farm Development*, New York: CRC Press, pp. 113–20.

Altieri, M. and Hecht, S. (eds) (1990) *Agroecology and Small Farm Development*, New York: CRC Press.

* Altieri, M. and Yurjevic, A. (1991) 'Influencias de las Relaciones Norte-Sur en la Investigación Agroecológica y Transferencia de Tecnología en América Latina: El Caso de CLADES'.

Annis, S. (1987) 'Can Small-Scale Development be a Large-Scale Policy? The Case of Latin America,' *World Development* 15 (supplement): 129–34.

Annis, S and Hakim, P. (eds) (1988) *Direct to the Poor: Grassroots Development in Latin America*, London: Lynne Reiner.

Anon, (1989) *Memoria. Seminario Taller Sobre Cooperación al Desarrollo*, La Paz, 8–9 June.

Arbab, F. (1988) *Non-Governmental Organizations: Report of a Learning Project*, Cali: CELATER.

REFERENCES

Arango, M. (1989) *Estudio sobre la Economía Campesina*, Misión de Estudios Agropecuarios, Medellín: Universidad de Antioquia/CIA.

Ayers, A. (1992) 'Development Assistance in San Julian, Eastern Bolivia: Conflicts and Complementarities', manuscript, Santa Cruz, Bolivia.

Barraclough, S. and Domike, A. (1966) 'Agrarian Structure in Seven Latin American Countries', *Land Economics*, 42(4): 391–424; reprinted in R. Stavenhagen (ed.) *Agrarian Problems and Peasant Movements in Latin America* Garden City. NY: Anchor Books, pp. 41–96.

Barraclough, S., van Buren, A., Gariazzo, A., Sundaram, A. and Utting, P. (1988) *Aid that Counts: The Western Contribution to Development and Survival in Nicaragua*, Amsterdam: Transnational Institute.

Barril, A. and Berdegué, J. (1988) 'Generación y Transferencia Tecnologica: la Exclusión de los Productores Campesinos', in GIA (ed.) *Gobierno Local y Participación Social: debate desde una perspectiva agraria*, pp. 204–39, Santiago: GIA.

Barry, T. and Preusch, B. (1988) *The Soft War: The Uses and Abuses of US Economic Aid in Central America*, New York: Groves Press.

Barsky, O. (1984) *La Reforma Agraria Ecuatoriana*, Quito: Corporación Editora Nacional.

—— (1990) *Políticas Agrarias en América Latina*, Santiago: CEDECO.

Barsky, O. and Llovet, I. (1983) *Pequeña Producción y Accumulación de Capital: los productores de papa de Carchi, Ecuador*, Publicación Miscelanea no. 369, PROTAAL documento no. 87, San Jose: IICA.

—— and —— (1986) 'Pequeña producción y accumulación de capital: los productores de papa de Carchi, Ecuador,' in M. Piñeiro and I. Llovet (1986) (eds): pp. 251–326.

Baumeister, E. (1987) 'Tendencias de la agricultura centroamericana en los años ochenta', *Cuadernos de Ciencias Sociales No. 7*, Facultad Latinoamericana de Ciencias Sociales, San Jose, Costa Rica.

Bebbington, A.J. (1989) 'Institutional Options and Multiple Sources of Agricultural Innovation: Evidence from an Ecuadorean Case Study', *ODI Agricultural Administration (Research and Extension) Network Paper 11*, Overseas Development Institute, London.

—— (1990) 'Farmer Knowledge, Institutional Resources and Sustainable Agricultural Strategies: A Case Study from the Eastern Slopes of the Peruvian Andes', *Bulletin of Latin American Research*, 9(2): 203–28.

——(1991) 'Sharecropping Agricultural Development: the Potential for GSO–Government Co-operation', *Grassroots Development* 15(2): 20–30.

—— (1992) 'Searching for an Indigenous Agricultural Development: Indian organizations and NGOs in the Central Andes of Ecuador', Centre of Latin American Studies, Cambridge University, *Working Paper No. 45*, Centre of Latin American Studies, Cambridge.

—— (forthcoming) 'Theory and Relevance in Indigenous Agriculture: Knowledge, Agency, Organization', in D. Booth (ed.) *New Directions in Social Development Research: Relevance Realism and Choice*, Cambridge: Cambridge University Press.

Bebbington, A.J. and Carney, J. (1990) 'Geographers in the International Agricultural Research Centers: Theoretical and Practical Considerations', *Annals of the Association of American Geographers* 80(1): 34–48.

Bebbington, A., Davies, P., Prager, M., Thiele, G. and Wadsworth, J. (1992) *Informe del Taller: Generación y Transferencia de Tecnología Agropecuaria: el Papel de las ONG y el Sector Público*, Santa Cruz: Centro de Investigación Agricola Tropical.

——, ——, ——, ——, ——, and —— (1992) 'From Protest to Proposals? Grassroots

Development Challenges to Ecuador's Indian Organizations', *Grassroots Development*, Fall 1992.
Bebbington, A.J., Carrasco, H., Peralbo, L., Ramón, G., Torres, V.H. and Trujillo, J. (1991) 'Evaluación del impacto generado por los proyectos de desarrollo de base auspiciados por la Fundación Inter-americana en el Ecuador', report to Inter-American Foundation, Rosslyn, USA.
Berdegué, J. (1990) 'NGOs and Farmers' Organizations in Research and Extension in Chile,' *Agricultural Administration (Research and Extension) Network Paper 19*, Overseas Development Institute, London.
Berdegué, J. and Nazif, I. (1988) *Sistemas de Producción Campesinos*, Santiago: Grupo de Investigaciones Agrarias.
Bernstein, R. (1983) *Beyond Objectivism and Relativism*, Oxford: Basil Blackwell.
Biggs, S. (1989) 'Resource-Poor Farmer Participation in Research: A Synthesis of Experiences From Nine National Agricultural Research Systems', *OFCOR Comparative Study Paper No. 3*, International Service for National Agricultural Research, The Hague.
—— (1990) 'A Multiple Source of Innovation Model of Agricultural Research and Technology Transfer', *World Development* 18(11): 1481–99.
Blanes, J. (1986) *Bolivia: Extrema Pobreza en la Agricultura. Causas y condiciones para el desarrollo*, Mimeo, La Paz.
Blauert, J. (1988) 'Autochthonous Development and Environmental Knowledge in Oaxaca, Mexico,' in P. Blaikie and T. Unwin (eds) *Environmental Crises in Developing Countries*, Monograph 5, Developing Areas Research Group, Institute of British Geographers.
Bobbio, N. (1987) *The Future of Democracy*, Cambridge: Polity Press.
*Bojanic, A. (1991) 'La Transferencia de Tecnología en Bolivia: La Marcha para llegar al Modelo de Usuarios Intermediarios'.
Booth, D. (ed.) (forthcoming) *New Directions in Social Development Research: Relevance, Realism and Choice.*
Breslin, P. (1991) 'Democracy in the Rest of the Americas', *Grassroots Development* 15(2): 3–7.
Brockett, C. (1988) *Land, Power, and Poverty: Agrarian Transformation and Political Conflict in Central America*, Winchester, Mass.: Allen & Unwin.
Browder, J.O. (ed.) (1989) *Fragile Lands of Latin America: Strategies for Sustainable Development*, Boulder: Westview Press.
Bulmer-Thomas, V. (1987) *The Political Economy of Central America since 1920*, Cambridge: Cambridge University Press.
Byerlee, D. (1987) 'Maintaining the Momentum in Post-Green Revolution Agriculture: A Micro-Level Perspective from Asia', *Michigan State University International Development Paper No. 10*, East Lansing, Michigan.
Byrnes, K. (1992) *A Cross Cutting Analysis of Agricultural Research, Extension and Education (AG REE) in AID-Assisted LAC Countries*, Washington: Chemonics International.
*CAAP (1991) 'Generación y Transferencia de Tecnología Agropecuaria. Sistematización de Experiencias en el CAAP'.
Caballero, J.M. (1984) 'Agriculture and the Peasantry under Industrialization Pressures: Lessons from the Peruvian Experience', *Latin American Research Review* 19(2): 3–42.
*Cardoso, V.H. (1991) 'Relación del Programa de Investigación en Producción del INIAP con las Organizaciones No Gubernamentales y Organizaciones Campesinas'.
Carroll, T. (1970) (1964) 'Land Reform as an Explosive Force in Latin America', in

R. Stavenhagen (ed.) *Agrarian Problems and Peasant Movements in Latin America*, Garden City, NY: Anchor Books, 1970, pp. 101–38.
—— (1992) *Intermediary NGOs: The Supporting Link in Grassroots Development*, West Hartford: Kumarian Press.
Carroll, T. Humphreys, D. and Scurrah, M. (1991) 'Grassroots Support Organizations in Peru', *Development in Practice* 1(2): 97–108.
CEPAL (1990) *Changing Production Patterns with Social Equity*, Santiago: Economic Commission for Latin America and the Caribbean.
CESA (1980) *Un Apoyo al Desarrollo Campesino*, Quito: Central Ecuatoriana de Servicios Agricolas.
*CESA (1991) 'La Relación de CESA con el Estado en la Generación y Transferencia de la Tecnología Agropecuaria'.
Chambers, R (1983) *Rural Development: Putting the Last First*, London: Longman.
—— (1987) 'Sustainable Livelihoods, Environment and Development: Putting Poor Rural People First', *Discussion Paper No. 240*, Institute of Development Studies, Brighton.
Chambers, R., Pacey, A. and Thrupp, L.A. (eds) (1989) *Farmer First: Farmer Innovation and Agricultural Research*, London: Intermediate Technology Publications.
*Chavez, J. (1991) 'Los Programas de Agricultura Ecológica y de Agro-Industria Alimentaria del Centro IDEAS'.
CIAT (1991) *CIAT in the 1990s and Beyond: A Strategic Plan*, Cali: Centro Internacional de Agricultura Tropical.
Clark, J. (1991) *Democratizing Development: The Role of Voluntary Organizations*, London: Earthscan.
Cleary, D. (1991) 'The 'Greening' of the Amazon', in D. Goodman and M. Redclift (eds) *Environment and Development in Latin America: The Politics of Sustainability*, Manchester: Manchester University Press, 1991, pp. 116–40.
Cleaves, P. and Scurrah, M. (1980) *Agriculture, Bureaucracy and the Military*, Ithaca: University of Cornell Press.
CONAIE (1989) *Nuestro Proceso Organizativo*, Quito: Confederación de Nacionalidades Indígenas del Ecuador.
Consultores SETA (1988) *Programa de Reforzamiento de la Investigación Agronómica para Granos Básicos en Centroamérica*, Panama: Consultores SETA.
COTESU-MACA-ILDIS (1990) *El impacto de la NPE en el sector agropecuario*, La Paz: COTESU-MACA-ILDIS.
Cotlear, D. (1986) 'Farmer Education and Farm Efficiency in Peru: The Role of Schooling, Extension Services and Migration', *Education and Training Series Discussion Paper*, Report No. EDT 49, The World Bank, Washington DC.
—— (1989a) 'The Effect of Education on Farm Productivity', *Journal of Development Planning* 19: 73–99.
—— (1989b) *El Desarrollo Campesino en los Andes. Cambio Tecnológico y Transformación Social en las Comunidades de la Sierra del Peru*, Lima: Instituto de Estudios Peruanos.
Coutu, A. and Gross, D. (1990) 'Technological Change and Sources of Change in Agricultural Production in Central America', paper presented at the Conference on the Transfer and Utilization of Agricultural Technology in Central America, Coronado, Costa Rica, 12–16 March.
Coutu, A. and O'Donnell, J. (1991) 'Agricultural Development Foundations: A Private Sector Innovation in Improving Agricultural Technology Systems', in W.M. Rivera and D.J. Gustafson (eds) *Agricultural Extension: Worldwide Institutional Evolution and Forces for Change*, Amsterdam: Elsevier, 1991, pp. 113–22.

de Janvry, A. (1981) *Land Reform and the Agrarian Question in Latin America*, Baltimore: Johns Hopkins University Press.

de Janvry, A. and Garcia, R. (1992) 'Rural Poverty and Environmental Degradation in Latin America', *Technical Issues in Rural Poverty Alleviation Staff Working Paper 1*, International Fund for Agricultural Development, Rome.

de Janvry, A, Marsh, R, Runsten, D, Sadoulet, E. and Zabin, C. (1989) 'Impacto de la Crisis en la Economía Campesina de América Latina y el Caribe', in F. Jordan (ed.) *La Economía Campesina: Crisis, Reactivación, Políticas*, San Jose: Instituto Interamericano de Cooperación para la Agricultura, 1989, pp. 91–205.

de Walt, B. (1988) 'Halfway There: Social Science in Agricultural Development and Social Science of Agricultural Development', *Human Organization* 47: 343–53.

Devé, F. (1987) 'Apoyo a la caracterización de los productores de granos básicos del Istmo Centroamericano', *Temas de Seguridad Alimentaria No. 4*, CADESCA/CCE, Panama.

Dunkerley, J. (1984) *Rebellion in the Veins*, London: Verso.

Duran, J. B. (1990) *Las nuevas instituciones de la sociedad civil. Impacto y tendencia de la co-operación internacional y las ONGs en el area rural de Bolivia*, La Paz: Huellas.

Echenique, J. (1991) 'Las Políticas en el Marco del Ajuste', manuscript, AGRARIA, Santiago.

Echenique, J. and Rolando, N. (1989) *La Pequeña Agricultura: Una Reserva de Potencialidades y Una Deuda Social*, Santiago: AGRARIA.

Echeverría, R. (1990) 'Inversiones públicas y privadas en la investigación sobre maiz en México y Guatemala', *Documento de Trabajo del Programa de Economía 90/03*, CIMMYT, Mexico.

Edwards, M. and Hulme, D. (eds) (1992) *Making a Difference? NGOs and Development in a Changing World*, London: Earthscan.

Eguren, A. (1990) 'Adjustment with Growth in Latin America', *Economic Development Institute Seminar Report No. 22*, World Bank, Washington, DC.

Engel, P. (1990) 'The Impact of Improved Institutional Coordination on Agricultural Performance: The Case of the Narino Highlands in Colombia', *Linkages Discussion Paper No. 4*, International Service for National Agricultural Research, The Hague.

Erickson, C. and Candler, K. (1989) 'Raised Fields and Sustainable Agriculture in the Lake Titicaca Basin of Peru,' in J.O. Browder (ed.) *Fragile Lands of Latin America: Strategies for Sustainable Development*, Boulder: Westview Press, 1989, pp. 230–48.

*Estrada, J.F. (1991) 'Analisis Evolutivo en el Proceso de Transferencia de Tecnología en el Instituto Colombiano Agropecuario ICA: Una Experiencia Institucional'.

Ekpere, J. and Idowu, I. (1989) *Managing the Links Between Research and Technology Transfer: the Case of the Agricultural Extension-Research Liaison Service in Nigeria*, The Hague: International Service for National Agricultural Research.

EXTIE-World Bank (1990) 'Strengthening the Bank's Work on Popular Participation: A Proposed Learning Process', mimeo, EXTIE-World Bank, Washington, DC.

FAO (1988) *Potentials for Agricultural and Rural Development in Latin America and the Caribbean. Annex V: Crops, Livestock Fisheries and Forestry*, Rome: Food and Agricultural Organization.

—— (1990) *Inventario de ONGs Vinculadas al Desarrollo Agropecuario y Rural* (by A. Peters and P. Mendez), La Paz: Food and Agriculture Organization.

Farrington, J. and Biggs, S.D. (1991) 'NGOs, Agricultural Technology and the Rural Poor', *Food Policy* 15(6): 479–91.

Farrington, J. and Lewis, D. (1993) (eds) *NGOs and the State in Asia: Rethinking Roles in Sustainable Agricultural Development*, London: Routledge.

Farrington, J. and Martin, A. (1988) 'Farmer Participatory Research: A Review of

Concepts and Practices', *ODI Agricultural Administration (Research and Extension) Discussion Paper No. 19*, Overseas Development Institute, London.
Farrington, J. and Mathena, S.D. (1991) 'Managing Agricultural Research for Fragile Environments: Amazon and Himalayan Case Studies', *ODI Occasional Paper No 11*, Overseas Development Institute, London.
Field, L. (1990) *Adaptación Socio-Económica Campesino Indígena de la Sierra Norte. Los Sistemas Agricolas en la Sierra Norte del Ecuador. Opciones para Analizar su Desarrollo. Informe Final*, Quito: Centro Andino de Acción Popular.
Figueroa, A. (1984) 'La via campesina al desarrollo rural en America Latina'. Figueroa, A. and Bolliger, F. (1985) *Productividad y Aprendizaje en el Medio Ambiente Rural. Informe Comparitivo*, Rio de Janeiro: ECIEL.
—— (1990) 'La Agricultura Campesina en América Latina: desafios para los 90', *Estudios Rurales Latinoamericanos* 13(1–2).
Fox, J. (ed.) (1990a) 'Editor's Introduction' in J. Fox (ed.) *The Challenge of Rural Democratisation: Perspectives from Latin America and the Phillipines*, London: Frank Cass, 1990, pp. 1–18.
—— (1990b) *The Challenge of Rural Democratization: Perspectives from Latin America and the Phillipines*, London: Frank Cass. This is also published as a special issue of *Journal of Development Studies* 26(4).
Freire, P. (1970) *Pedagogy of the Oppressed*, New York: Seabury Press.
Friedmann, J. (1992) *Empowerment: The Politics of Alternative Development*, Oxford: Basil Blackwell.
*FUNDAEC (1991) 'FUNDAEC y el Sector Público: Un Enfoque de Integración entre las ONG y Organizaciones Gubernamentales para la Generación y Transferencia de Tecnologías Agropecuarias'.
*Garcia, R., Warmenbol, K and Matsuzaki, S. (1991) 'Transferencia de Tecnología y Desarrollo Comunitario: Experiencia de CIPCA en el Departamento de Santa Cruz, Bolivia'.
Giddens, A. (1984) *The Constitution of Society*, Cambridge: Polity Press.
Gomez, S. and Echenique, J. (1988) *La Agricultura Chilena: las dos Caras de la Modernización*, Santiago: FLACSO.
*Gonzalez, R. (1991) 'Generación y Transferencia de Tecnología Agropecuaria: el Papel de las ONG y el Sector Público: El Caso del Secretariado Diocesano de Pastoral Social 'SEPAS' San Gil, Santander, Colombia'.
*Gonzalez, W. (1991) 'PROCADE: Un Análisis de Su Rol Coordinador de Actividades Interinstitucionales y de Relación con el Sector Público Agropecuario'.
Goodman, D. and Hall, A. (1990) (eds) *The Future of Amazonia: Destruction or Sustainable Development*, London: Macmillan.
Goodman D. and Redclift, M. (eds) (1991) *Environment and Development in Latin America: The Politics of Sustainability*, Manchester: Manchester University Press.
Griffin, K. (1975) *The Political Economy of Agrarian Change: An Essay on the Green Revolution*, London: Macmillan.
Griffin, K. and Knight, J. (1989) 'Human Development: The Case for Renewed Emphasis', *Journal of Development Planning* 19: 9–40.
Grindle, M S. (1986) *State and Countryside: Development Policy and Agrarian Politics in Latin America*, Baltimore: Johns Hopkins University Press.
Grupo Esquel (1989) 'Las Políticas de Desarrollo Rural en América Latina: balance y perspectivas', in F. Jordan (ed.) *La Economía Campesina: Crisis, Reactivación y Desarrollo*, San Jose: Instituto Interamericano de Cooperación para la Agricultura, 1989, pp. 5–90.
Grupo Regional de Fortalecimiento Institucional (1992) *Análisis de los sistemas estatales encargados de generar y transferir tecnología en granos básicos, documento*

síntesis de los países centroamericanos, San Jose, Costa Rica: Programa Regional de Reforzamiento de la Investigación Agronómica sobre los Granos en Centroamerica. Convenio CORECA–CEE–IICA.

Guggenheim, S. (1989) 'Recent Trends in Small Farmer Input Use in Latin America', in D. Groenfeldt and J.L. Moock (eds) *Social Science Perspectives on Managing Agricultural Technology*, Colombo: International Irrigation Management Institute, pp. 155–67.

Habermas, J. (1984) *The Theory of Communicative Action*, Cambridge: Polity Press.

Haney E.B. and Haney W.G. (1989) 'The Agrarian Transition in Highland Ecuador: From Precapitalism to Agrarian Capitalism in Chimborazo', in W. Thiesenhusen (ed.) *Searching for Agrarian Reform in Latin America*, London: Unwin Hyman, 1989, pp. 70–91.

Hayami, Y. and Ruttan, V. (1985) *Agricultural Development: An International Perspective*, rev. ed, Baltimore: Johns Hopkins University Press.

Healey, J. and Robinson, M. (1992) *Democracy, Governance and Economic Policy: Sub-Saharan Africa in Comparative Perspective*, London: Overseas Development Institute.

Healy, K. (1988) 'From Field to Factory: Vertical Integration in Bolivia', *Grassroots Development*, 11(2): 2–11.

Hecht, S. (1985) 'Environment, Development and Politics: The Livestock Sector in Eastern Amazon', *World Development* 13(6): 663–85.

Heller, P. (1990) 'Fund-Supported Adjustment Programs and the Poor', in IMF and World Bank (eds) *The Path to Reform: Issues and Experiences*, Washington, DC: International Monetary Fund, 1990, pp. 9–13.

Hewitt de Alcantará, C. (1976) *Modernizing Mexican Agriculture*, Geneva: United Nations Research in Social Development.

Hildebrand, P. and Ruano, S. (1982) *El Sondeo*, Guatemala: Instituto de Ciencia y Tecnología Agricolas.

Hirschmann, A. (1984) *Getting Ahead Collectively: Grassroots Experiences in Latin America*, Oxford: Pergamon.

Horton, D., Tardieu, F., Benavides, M., Tomassini, L. and Accotino, P. (1980) 'Tecnología de la Produccion de Papa en el Valle del Mantaro, Perú. Resultados de una Encuesta Agro-Económica de Visita Multiple', *Documento de Trabajo 1980-1*, Departamento de Ciencias Sociales, Centro Internacional de la Papa, Lima.

IDB (1986) *Economic and Social Progress in Latin America. 1986 Report*, Washington, DC: Inter-American Development Bank.

—— (1991) 'Report on The First IDB–European NGO Consultation, Paris September 13th, 1991', mimeo, Inter-American Development Bank, Paris.

IFAD (1991) *Republic of Peru: Strengthening of Agricultural Extension Services for Peasant Communities in the Highlands of Peru: Project Brief*, Rome: International Fund For Agricultural Development.

IICA (1987) *Memoria del Taller sobre Investigación y Tecnología Agropecuaria para el Pequeño Productor: ONGs-Procisur-Prociandino*, Santa Cruz de la Sierra, 17–19 November 1987, San Jose: Instituto Interamericano de Cooperación para la Agricultura.

—— (1991a) *PRONADER. Estudio de Factibilidad: Resumen General de Proyecto*. Quito: Instituto Interamericano de Cooperación para la Agricultura.

—— (1991b) *Regional Overview of Food Security in Latin America and the Caribbean with a Focus on Agricultural Research and Technology Transfer*, Programa de Generacion y Transferencia de Tecnologia, San Jose: Instituto Interamericano de Cooperación para la Agricultura.

IICA-CENTA (Inter-American Institute for Cooperation in Agriculture, Basic Grains

Program, Centre for Agricultural Technology) (1992) *Seminario - Taller: El Papel de las ONG's, el estado y los organismos de cooperación internacional en la generación y transferencia de tecnología agropecuaria*, San Salvador: IICA-CENTA.

ILEIA (1989) *ILEIA Newsletter*, Special Issue on Participatory Technology Development.

INDAP (1992) *Bases para la Inscripción en el Registro Nacional de Consultores de Transferencia Tecnológica Modalidad Cofinanciada Temporada 1992–1993*, Santiago: Instituto de Desarrollo Agropecuario.

Instituto para el Desarrollo Económico y Social de América Central (IDESAC), Alianza para el Desarrollo Juvenil Comunitario (ALIANZA), Proyecto de Desarrollo Santiago (PRODESA), Comité Central Menonita - Tecnología Apropiada (CCM-TA) (1989) *Directorio de instancias que promueven tecnología apropiada agropecuaria*, Guatemala: IDESAC.

IRENA/MIRENEM (Instituto Nicaragüense de Recursos Naturales y del Ambiente Ministerio de Recursos Naturales, Energía y Minas de Costa Rica) (1991) *Marco conceptual y plan de acción para el desarrollo del sistema internacional de areas protegidas para la paz, Si-a-Paz*, Managua: IRENA/MIRENEM.

ISNAR. (1989a) *Reforzamiento del Instituto Nacional de Investigaciones Agropecuarias: Base para un Sistema Nacional de Investigación Agropecuaria*, The Hague: International Service for National Agricultural Research.

—— (1989b) *Fortalecimiento del Sistema de Investigación y Transferencia de Tecnología Agropecuaria en Bolivia*, The Hague: International Service for National Agricultural Research.

—— (1992) *Summary of Agricultural Research Policy: International Quantitative Perspectives*, The Hague: International Service for National Agricultural Research.

Jennings, B.H. (1988) *Foundations of International Agricultural Research: Science and Politics in Mexican Agriculture*, Boulder: Westview Press.

Johnston, B. and Kilby, P. (1975) *Agriculture and Structural Transformation: Economic Structure in Late Developing Countries*, Oxford: Oxford University Press.

Joint Ecuadorian/NC State University Subcommittee (1987) *Reorientation of the Agricultural Sector: A Strategy to Accelerate Application of Science to Increase Agricultural Productivity in Ecuador*, Quito: Ministerio de Agricultura y Ganadería.

Jordan, F. (ed.) (1989) *La Economía Campesina: Crisis, Reactivación y Desarrollo*, San Jose: Instituto Interamericano de Cooperación para la Agricultura.

Jordan, F., de Miranda, C., Reuben, W. and Sepulveda, S. (1989) 'La Economía Campesina en la Reactivación y el Desarrollo Agropecuario', in F. Jordan (ed.) *La Economía Campesina: Crisis, Reactivación y Desarrollo*, San Jose: Instituto Interamericano de Cooperación para la Agricultura, 1989, pp. 207–88.

Kaimowitz, D. (1989) 'Placing Agricultural Research in one Organization: Two Experiences from Colombia', *Linkages Discussion Paper No. 3*, International Service for National Agricultural Research, The Hague.

Kaimowitz, D. (ed.) (1990) *Making the Link: Agricultural Research and Technology Transfer in Developing Countries*, Boulder: Westview and International Service for National Agricultural Research.

*—— (1991) 'El Papel de las ONG en el Sistema Latinoamericano de Generación y Transferencia de Tecnología Agropecuaria'.

—— (1992) 'El apoyo tecnológico necesario para promover las exportaciones agrícolas no tradicionales en America Central', *Serie Documentos de Programas #30*, San Jose: Instituto Interamericano de Cooperación para la Agricultura.

Kaimowitz, D., Erazo, D., Mejía, M. and Navarro, A. (1992) *Las organizaciones privadas de desarrollo y la transferencia de tecnología en el agro hondureño*, Tegucigalpa: Instituto Interamericano de Cooperación para la Agricultura.

Kaimowitz, D., Snyder, M. and Engel, P. (1990) 'A Conceptual Framework for Studying the Links between Agricultural Research and Technology Transfer in Developing Countries', in D. Kaimowitz (ed.) *Making the Link: Agricultural Research and Technology Transfer in Developing Countries*, Boulder: Westview and International Service for National Agricultural Research.

Kay, C. 1982. 'Achievements and Contradictions of the Peruvian Agrarian Reform', *Journal of Development Studies* 18(2): 141–70.

Kean, S. and Singogo, L. (1990) *Bridging the Gap Between Research and Extension in Zambia: The Incorporation of Research-Extension Liaison Officers into the Adaptive Research Planning Team* (OFCOR Case Study Number 1), The Hague: International Service for National Agricultural Research.

Kleemeyer, C. (1991) 'What is Grassroots Development?' *Grassroots Development* 15(1): 38–9.

Kohl, B. (1991) 'Protected Horticultural Systems in the Bolivian Andes: A Case Study of NGOs and Inappropriate Technology', *Agricultural Administration (Research and Extension) Network Paper 29*, Overseas Development Institute, London.

*Kopp, A. and Domingo, T. (1991) 'Tecnologías de Conservación en el Trópico: CESA, Bolivia'.

Krantz, L., Salick, J., Walter, S. and Norheim, T. (1992) *Conservation for Sustainable Development Project in Central America (OLAFO), A Review of Phase I: 1989–1992*, Final Report.

Krueger, A. (1974) 'The Political Economy of the Rent-Seeking Society,' *American Economic Review* 164(3): 291–303.

Landim, L. (1987) 'Non-Governmental Organizations in Latin America,' in A. Drabek (ed.) 'Development Alternatives: The Challenge for NGOs', *World Development* 15 (supplement): 29–38.

Landsberger, H. and Hewitt, C. (1970) 'Ten Sources of Weakness and Cleavage in Latin American Peasant Movements,' in R. Stavenhagen (ed.) *Agrarian Problems and Peasant Movements in Latin America*, Garden City, NY: Anchor Books, 1970, pp. 559–83.

Lehmann, A.D. (1978) 'The Death of Land Reform: A Polemic', *World Development* 6(3): 339–45.

—— (1982) 'Beyond Lenin and Chayanov: New paths of agrarian capitalism,' *Journal of Development Economics* 11:133–61.

—— (1984) *Economic Development and Social Differentiation in the Andean Peasant Economy*, Cambridge: University of Cambridge Department of Applied Economics.

—— (1990) *Democracy and Development in Latin America: Economics, Politics and Religion in the Postwar Period*, Cambridge: Polity Press.

Levitt, A., Jr and Picado, S. (1989) 'Pobreza, conflicto y esperanza, Informe de la Comisión Sanford para la Recuperación y el Desarrollo de Centro América', *Panorama Centroamericano*, 19–20 (January–April), Instituto de Estudios Políticos, Guatemala.

Lindarte, E. (1990) 'Technological Institutions in the Region: Evolution and Current State', paper presented at the Conference on the Transfer and Utilization of Agricultural Technology in Central America, Coronado, Costa Rica, 12–16 March.

Lipton, M. and Longhurst, R. (1989) *New Seeds and Poor People*, London: Unwin Hyman.

Long, N. (1988) 'Sociological Perspectives on Agrarian Development and State Intervention,' A. Hall and J. Midgley (eds) *Development Policies: Sociological Perspectives*, Manchester: Manchester University Press.

Long, N. (1990) 'From Paradigm Lost to Paradigm Regained? The Case for an Actor-oriented Sociology of Development', *European Review of Latin American and Caribbean Studies* 49: 3–32.

Long, N. and van der Ploeg, J. (forthcoming) 'Reflections on the Actor-Oriented Approach to Social Development Research: Towards a New Concept of Structure,' in D. Booth (ed.) *New Directions in Social Development Research: Relevance, Realism and Choice.*

Long, N. and Roberts, B. (1978) *Peasant Cooperation and Capitalist Expansion in Central Peru*, Austin: Unversity of Texas.

Loveman, B. (1991) 'NGOs and the Transition to Democracy in Chile,' *Grassroots Development* 15(2): 8–19.

Machado, A. (1986) *El Desarrollo Agricola y Compesino en Colombia*, Documento de Trabajo, Bogota: Centro Internacional de Investigaciones para el Desarrollo.

Martínez Noguiera, R. (1990) 'The Effect of Changes in State Policy and Organizations on Agricultural Research and Extension Links: A Latin American Perspective', in D. Kaimowitz (ed.) *Making the Link: Agricultural Research and Technology Transfer in Developing Countries*, Boulder: Westview and International Service for National Agricultural Research, 1990, pp. 75–108.

Matheu, R. (1991) *El papel de las organisaciones no gubernamentales de desarrollo y de las cooperativas en la generación y transferencia de tecnología agropecuaria en Centroamerica, caso: Guatemala*, Guatemala: Programa Regional de Reforzamiento de la Investigación Agronómica sobre los Granos en Centroamerica, Convenio CORECA–CEE–IICA.

Merrill-Sands, D. (1989) 'Introduction to the ISNAR Study on Organization and Management of On-Farm Client Oriented Research (OFCOR)', in S. Biggs, 'Resource-Poor Farmer Participation in Research: A Synthesis of Experiences from Nine National Agricultural Research Systems', *OFCOR Comparative Study Paper No. 3*, International Service for National Agricultural Research, The Hague.

Merrill-Sands, D. and Kaimowitz, D. (1990) *The Technology Triangle: Linking Farmers, Technology Transfer Agents and Agricultural Researchers*, The Hague: International Service for National Agricultural Research.

Meyer, C. (1992) 'A Step Back as Donors Shift Institution Building from the Public to the "Private" Sector', *World Development* 20(8): 1115–26.

Milz, J. (1990) 'La Producción de Cacao Biológico. Un Compromiso entre Ecología y Economía Campesina,' in COTESU (ed.) *Desarrollo y Medio Ambiente*, La Paz: COTESU, pp. 20–21.

Morales, J.A. (1990) 'Impacto de los ajustes estructurales en la agricultura campesina boliviana', in COTESU-MACA-ILDIS (ed.) *El impacto de la NPE en el sector agropecuario*, La Paz: COTESU-MACA-ILDIS, 1990, pp. 9–69.

Morgan, M. (1991) 'Stretching the Development Dollar: The Potential for Scaling-Up', *Grassroots Development* 14(1): 2–11.

Muñoz, S. and Dandler, J. (1986) *Importancia del Campesinado en los Sistemas de Comercialización de Alimentos en Bolivia*, La Paz: Ministerio de Planeación y Coordinación/UNICEF.

Newman, J, Jorgenson, S. and Pradhan, M. (1991) 'Workers' Benefits from Bolivia's Emergency Social Fund', *Living Standards Measurement Study Working Paper No.* 77, World Bank, Washington, DC.

Nunberg, B. (1988) *Public Sector Management Issues in Structural Adjustment Lending* (*World Bank Discussion Paper* 99), Washington, DC: World Bank Publications.

Oakley, P. (1991) *Projects with People: The Practice of Participation in Rural Development*, Geneva: International Labour Office.

O'Brien, J. (1991) 'How Can Donors Best Support NGO Consortia?' *Grassroots Development* 15(2): 38–40.

ODI (1992) 'Aid and Political Reform', *ODI Briefing Paper*, January 1991, Overseas Development Institute, London.

Ortiz, R., Ruano S., Juarez, H., Olivet, F., and Meneses, A. (1991) 'A New Model for Technology Transfer in Guatemala: Closing the Gap Between Research and Extension', *OFCOR Discussion Paper No. 2*, International Service for National Agricultural Research, The Hague.

Padrón, M. (1982) *Cooperación al Desarrollo y Movimiento Popular: las Associaciones Privadas de Desarrollo*, Lima: DESCO.

Palmieri, V. (1990) 'Efectos de los cambios estructurales en el Ministerio de Agricultura y Ganadería de Costa Rica, sobre la relación entre investigación y transferencia de tecnología en maíz', *Linkage Discussion Paper No. 7*, International Service for National Agricultural Research, The Hague.

Pardey, P.G., Roseboom, J. and Anderson, J.R. (1991) *Agricultural Research Policy: International Quantitative Perspectives*, Cambridge, Cambridge University Press.

Piñeiro, M. (1985) 'The Development of the Private Sector in Agricultural Research: Implications for Public Research Institutions', *PROAGRO Paper No. 10*, International Service for National Agricultural Research, The Hague.

Piñeiro, M. and Llovet, I. (eds) (1986) *Transición Tecnológica y Differenciación Social*, San Jose: Instituto Interamericano para la Cooperación en la Agricultura.

Piñeiro, M. and Trigo, E. (eds) (1983) *Technical Change and Social Conflict in Agriculture: Latin American Perspectives*, Boulder: Westview Press.

Piñeiro, M., Trigo, E. and Ardila, J. (1985) *Organización de la Investigación Agropecuaria en América Latina*, San Jose: Instituto Interamericano para la Cooperación en la Agricultura.

Pray, C. and Echeverria, R. (1990) 'Private Sector Agricultural Research and Technology Transfer Links in Developing Countries', in D. Kaimowitz (ed.) *Making the Link: Agricultural Research and Technology Transfer in Developing Countries*, Boulder: Westview and International Service for National Agricultural Research, pp. 197–226.

PREIS (Programa Regional de Investigación sobre El Salvador) (1992) 'La sostenibilidad en las diversas propuestas para el sector agropecuario, la dimensin institucional', *Cuaderno de Trabajo No. 9*, San Salvador.

Reid, M. (1974) *Peru: Paths to Poverty*, London: Latin American Bureau.

Reilley, C. (1992) 'Making Government Safe for Democracy', *Grassroots Development* 16(1): 43–4.

Reuben, W. (1985) *Indigenous Agricultural Revolution*, London: Hutchinson.

—— (1991) 'El papel de las ONGs en la cooperación europea hacia Centroamérica', in W. Rueben and G. Van Oord (eds) *Mas allá del ajuste, la contribución europea al desarrollo democrático y duradero de las economías centroamericanas*, San Jose: Editorial DEI, pp. 337–92.

—— (1992) 'Entre la democratización y el ajuste estructural: nuevos protagonismos y el papel de las ONG', in W. Reuben and E. Guadamuz (eds) *El papel de las ONG (compilación)*, San Jose: Centro de Capacitación para el Desarrollo, pp. 55–66. (*Cuadernos de capacitación No. 1*).

Rhoades, R.E. (1984) *Breaking New Ground: Agricultural Anthropology*, Lima: Centro Internacional de la Papa.

Rhoades, R.E. and Bebbington, A. (1993) 'Farmers who Experiment. An Untapped Resource for Agricultural Research and Development,' in *Indigenous Knowledge Systems: The Cultural Dimension of Development*, D.M. Warren, D. Brokensha and L. Jan Slikkerveer (eds), London: Kegan Paul International.

Ribe, H., Carvalho, S., Liebenthal, R., Nicholas, P. and Zuckerman, E. (1990) 'How

Adjustment Programs Can Help the Poor: The World Bank's Experience', *World Bank Discussion Paper No. 71*, Washington, DC.
Richards, P. (1990). 'Indigenous Approaches to Rural Development: The Agrarian Populist Tradition in West Africa', in M. Altieri and S. Hecht (eds) *Agroecology and Small Farm Development*, New York: CRC Press, pp. 105–11.
—— (1985) *Indigenous Agricultural Revolution*, London: Hutchinson.
Riddell, R. and Robinson, M. (1993) 'Working with the Poor: NGOs and Rural Poverty Alleviation', manuscript, Overseas Development Institute, London.
Rigg, J. (1989) 'The New Rice Technology and Agrarian Change: Guilt by Association?' *Progress in Human Geography* 13(2): 374–99.
Ritchey-Vance, M. (1991) *The Art of Association: NGOs and Civil Society in Colombia*, Washington, DC: Inter-American Foundation.
Rivera-Cusicanqui, S. (1990) 'Liberal Democracy and Ayllu Democracy in Bolivia: The Case of Northern Potosi', in J. Fox (ed.) *The Challenge of Rural Democratization: Perspectives from Latin America and the Phillipines*, London: Frank Cass, 1990, pp. 97–121.
Rivera W.M. and Gustafson, D.J. (eds) (1991) *Agricultural Extension: Worldwide Institutional Evolution and Forces for Change*, Amsterdam: Elsevier.
Rodriguez, R. (1991) *El papel de los organismos no gubernamentales y federaciones de cooperativas en la generación y transferencia de tecnología agropecuaria, caso: El Salvador*, San Salvador: Programa Regional de Reforzamiento de la Investigación Agronómica sobre los Granos en Centroamerica. Convenio CORECA-CEE-IICA.
Rodriguez, R., Navarro, L., and Sanchez, R. (1984) *El desarrollo y transferencia de tecnología agrícola en El Salvador, Experiencia en el cultivo de maíz y sorgo*, San Salvador: CENTA. (*Boletín Divulgativo 21*).
Ruttan, V. (1991) 'Challenges to Agricultural Research in the 21st Century,' in P.G. Pardey, J. Roseboom and J.R. Anderson (eds) *Agricultural Research Policy: International Quantitative Perspectives*, Cambridge: Cambridge University Press, pp. 399–411.
Sandoval Z.G. (1988) *Organizaciones no gubernamentales de desarrollo en América Latina y el Caribe*, La Paz: CEBEMO-UNITAS.
—— (1990) 'Informe sobre el rol de las ONGs en Bolivia: 1990' Ponencia al taller 'El desarrollo en Bolivia y las ONGs', La Paz y Santa Cruz, 11–12 de octubre 1990.
—— (1991) 'Origen y desarrollo de las ONG's', in COTESU (ed.) *Estado y ONG's: De la Competencia a la Complementariedad?* La Paz: COTESU.
Schuh, G.E. (1992) 'Sustainability, Marginal Areas and Agricultural Research', *Technical Issues in Rural Poverty Alleviation Staff Working Paper 4*, International Fund for Agricultural Development, Rome.
Schultz, T. (1964) *Transforming Traditional Agriculture*, New Haven: Yale University Press.
Schultz, T. (1975) 'The Value of the Ability to Deal with Disequilibria', *Journal of Economic Literature* 13: 827–46.
Scott, J.C. (1985) *Weapons of the Weak: Everyday Forms of Peasant Resistance*, London: Yale University Press.
Selowsky, M. (1990) 'Stages in the Recovery of Latin America's Growth', in IMF and World Bank (eds) *The Path to Reform: Issues and Experiences*, Washington, DC: International Monetary Fund, pp. 61–4.
Sims, H. and Leonard, D. (1990) 'The Political Economy of the Development and Transfer of Agricultural Technologies', in D. Kaimowitz (ed.) *Making the Link: Agricultural Research and Technology Transfer in Developing Countries*, Boulder: Westview and International Service for National Agricultural Research, 1990, pp. 43–74.

Slater, D. (ed.) (1985) *New Social Movements and the State in Latin America*, Amsterdam: CEDLA.

Smith, N., Alvim, P., Serrao, A., Homma, A. and Falesi, I. (1991) 'Amazonia', manuscript, PROCEZ, Graduate School of Geography, Clark University.

Sollis, P. (1991). 'Multilateral Agencies and NGOs in the Context of Policy Reform', paper presented to the Conference on Changing US and Multilateral Policy Toward Central America, 10–12 June 1992, Washington, DC.

Soliz, R., Espinosa, P. and Cardoso, V.H. (1989) *Ecuador: Organización y Manejo de la Investigación en Finca en el Instituto Nacional de Investigaciónes Agropecuarias (INIAP)* (OFCOR Case Study No. 7), The Hague: International Service for National Agricultural Research.

Sotomayor, O. (1991) 'GIA and the New Chilean Public Sector: The Dilemmas of Successful NGO Influence over the State', *Agricultural Administration (Research and Extension) Network Paper 30*, Overseas Development Institute, London.

Southgate, D. (1991) 'Tropical Deforestation and Agricultural Development in Latin America', *World Bank Environment Department Policy Research Division Working Paper No. 1991–20*, The World, Washington, DC.

Stavenhagen, R. (1970) (ed.) *Agrarian Problems and Peasant Movements in Latin America*, Garden City, NY: Anchor Books.

Sthaler-Sholk, R. and Spoor, M. (1990) *Política macroeconmica y sus efectos en la agricultura y la seguridad alimentaria, Nicaragua*, Panama: CADESCA/CCE.

Stremlau, C. (1987) 'NGO Coordinating Bodies in Africa, Asia and Latin America', *World Development* 15 (supplement): 213–26.

Tendler, J., Healey, K. and O'Laughlin, C.N. (1988) 'What to Think About Cooperatives: A Guide from Bolivia', in S. Annis and P. Hakim (eds) *Direct to the Poor: Grassroots Development in Latin American*, London: Lynne Reiner, p. 85–116.

Thiele, G, Davies, P. and Farrington, J. (1988) 'Strength in Diversity: Innovation in Agricultural Technology Development in Eastern Bolivia', *ODI Agricultural Administration (Research and Extension) Network Paper 1*, Overseas Development Institute, London.

Thiesenhusen, W. (1989a). 'Introduction: Searching for Agrarian Reform in Latin America', in W. Thiesenhusen (ed.) *Searching for Agrarian Reform in Latin America*, London: Unwin Hyman, 1989, pp. 1–41.

—— (ed.) (1989b) *Searching for Agrarian Reform in Latin America*, London: Unwin Hyman.

*Torres, R. (1991) 'Organizaciones Nongubernamentales y Sector Público en el Peru: La Experiencia de la Comisión Coordinadora de Tecnología Andina'.

Treacy, J. (1989) 'Agricultural Terraces in Peru's Colca Valley: Promises and Problems of an Ancient Technology', in J. Browder (ed.) *Fragile Lands of Latin America: Strategies for Sustainable Development*, Boulder: Westview Press, 1989, pp. 209–29.

*Trujillo, G. (1991) 'Investigación y Extensión por El Ceibo en el Alto Beni'.

Turner, B. and Harrison, P. (1983) (eds) *Pulltrouser Swamp: Ancient Maya Habitat, Agriculture and Settlement in Northern Belize*, Austin: University of Texas Press.

UNICEF (1990) *State of the World's Children, 1990*, Oxford: Oxford University Press.

—— (1992) *State of the World's Children 1992*, Oxford: Oxford University Press.

United Nations (1988) *World Demographic Estimates and Projections, 1950–2025*, New York: United Nations.

Uquillas, J. (1992) 'Research and Extension Practice and Rural People's Agroforestry Knowledge in Ecuadorian Amazonia', manuscript, FUNDAGRO, Quito.

Velasco, J. and Barrios, F. (1989) "Perspectivas, priorización y objectivos de las IPDS en su relacionamiento con las agencias de co-operación', Ponencia al seminario, 'Co-operación al Desarrollo', La Paz, 1990.

*Veléz, R. and Thiele, G. (1991) 'Primeras Experiencias con un Nuevo Modelo de Transferencia de Tecnología'.
Vessuri, H. (1990) 'O Inventamos o Erramos: The Power of Science in Latin America', *World Development* 18(11): 1543–53.
Werter, F. (1991) *Hacia un Modelo de Preextensión en Bolivia (Documento de Discusión)*, La Paz: DHV Consultants.
Wilken, G. (1989) 'Transferring Traditional Technology: A Bottom-Up Approach for Fragile Lands', in J. Browder (ed.) *Fragile Lands of Latin America: Strategies for Sustainable Development*, Boulder: Westview Press, 1989, pp. 44–57.
Williams, A. (1990) 'A Growing Role for NGOs in Development', *Finance and Development* December 1991: 31–3.
Williams, R. (1986) *Export Agriculture and the Crisis in Central America*, Chapel Hill: University of North Carolina Press.
Wilson, M. (1991) 'Reducing the Costs of Public Extension Services: Initiatives in Latin America', in W.M. Rivera and D.J. Gustafson (eds) *Agricultural Extension: Worldwide Institutional Evolution and Forces for Change*, Amsterdam: Elsevier, 1991, pp. 13–20.
Woodgate, G. (1991) 'Agroecological Possibilities and Organisational Limits: Some Initial Impressions from a Mexican Case Study', in D. Goodman and M. Redclift (eds) *Environment and Development in Latin America: The Politics of Sustainability*, Manchester: Manchester University Press, 1991, pp. 184–204.
World Bank (1990a) *Social Investment in Guatemala, El Salvador, and Honduras*, report on the Workshop on Poverty Alleviation, Basic Social Services and Social Investment Funds, Paris, 29–30 June 1990, Washington, DC: World Bank.
—— (1990b) *Social Indicators of Development*, Baltimore: Johns Hopkins University Press.
—— (1991a) *El Banco Mundial y las organizaciónes no gubernamentales*, Washington, DC: World Bank.
—— (1991b) *World Development Report 1991: The Challenge of Development*, Oxford: Oxford University Press.
—— (1992) *World Development Report, 1992*, Washington, DC: World Bank.
—— (n.d.1) 'The Bank and NGOs: Recent Experience and Emerging Trends', draft, Washington DC.
—— (n.d.2) 'Cooperation between the World Bank and NGOs: 1990 Progress Report', discussion draft, Washington, DC.
Wurgaft, J. (1992) 'Social Investment Funds and Economic Restructuring in Latin America', *International Labour Review* 131(1): 35–44.
*Zeller, T. (1991) 'COTESU y las ONG'.
Zuquilanda, M. (1988) *Tradición y Actualidad en el Agro Serrano*, Quito: CEDIME.

INDEX